Thermodynamics

Thermodynamics

H. J. Kreuzer

Dalhousie University, Canada

Isaac Tamblyn

Lawrence Berkeley National Laboratory, USA

World Scientific

NEW JERSEY · LONDON · SINGAPORE · BEIJING · SHANGHAI · HONG KONG · TAIPEI · CHENNAI

- No subject or topic should be introduced as "well known" but succinctly and precisely.
- No step in a derivation should be labeled "obvious".
- Because the pioneers of thermodynamics were frequently "amateurs" and had very colorful careers, brief historical remarks will be presented including amusing anecdotes and pictures.
- As environmental concerns must at least in part be addressed by thermodynamic methods, we will frequently lean towards those concerns for examples, such as investigating the thermodynamics of hurricanes.
- We address other topics not typically found in undergraduate textbooks, such as thermodynamics of surfaces and polymers.

The material presented in this book has been used by one of us (HJK) for many years in a one-semester course given at the second, third, or fourth year level. The prerequisites are consequently minimal: a working knowledge of functions of many variables and partial derivatives; a primer on the necessary mathematics is given as an Appendix. Concepts and facts from first and second year physics classes should also be familiar.

At this point, we would like to mention that the two authors have approached the creation of this text from essentially opposite ends of their academic careers. HJK is a senior scientist who has taught the subject for a number of years. Conversely, IT (once a student of HJK) has only recently completed his formal academic training. This collaboration has resulted in a unique view on the topic: one that has the breadth of experience that can only be earned with time, yet is still very connected and aware of the perspective (and limitations) of a student new to the field. We have therefore focused on producing a narrative that pays careful attention to address both the conceptual and technical hurdles that students typically face. Furthermore, we have arranged the material in a manner consistent with how it is best communicated in a classroom.

Important concepts are addressed more than once. We first introduce the essence of an argument and in subsequent discussion it is refined and analyzed. This ensures that details of an argument, while crucial for a complete understanding, are not so overwhelming that key physics cannot be distilled. This approach is in keeping with the spirt in which thermodynamics was constructed. As the reader shall see, this was not a theory that was written down by one individual at one particular time. Like students facing a new subject, it was struggled with over time.

Our aim and hope for this endeavour will be achieved if the students gain a working knowledge of thermodynamics, and if ultimately they appreciate the intellectual beauty of thermodynamics as expressed in the quotes by Einstein and Eddington.

H.J. Kreuzer and I. Tamblyn
April 2010
Halifax and Berkeley

Contents

Chapter 1

Scope of Thermodynamics

The aim of physics is to explain the static and dynamic properties of matter and energy in space and time. Matter is made up of atoms which in turn are composed of electrons and nuclei, the latter being conglomerates of elementary particles, mostly protons and neutrons. From X-ray scattering we know that in crystalline solids, as an example, atoms are spaced a few angstroms (1 Å=10^{-10} m) apart. In conjunction with the fact that solids are fairly incompressible, one concludes that the size of atoms is a few Å. This implies that a macroscopic piece of matter consists of a large number of atoms, typically 10^{23} per cm^3 in solids and, as another example of a macroscopic object, about 10^{19} per cm^3 in a gas at atmospheric pressure and room temperature. The question then arises: must we know everything about the motion of this tremendous number of building blocks to understand properties of macroscopic objects? The answer is no, as thermodynamics provides a theoretical framework to deal with the properties of **macroscopic** systems. Its structure is so general that it can deal not only with physical systems but also with biology, environmental issues, and even financial systems. What is new in thermodynamics, as compared to mechanics or electromagnetism, is the fact that thermal effects are accounted for by including temperature and entropy among its variables and laws. One might say that thermodynamics is the science of heat transfer.

From its beginnings, thermodynamics was developed and used in the physical sciences, chemistry, and physics. Early investigations of thermodynamic principles were also undertaken in physiology and the medical sciences. And with the rise of engineering in the Industrial Age of the nineteenth century, thermodynamics became a cornerstone in practical issues of building machines and running factories. In recent decades new applications for thermodynamic principles have been explored in such diverse fields as economics, manufacturing, financial markets, and ecological sciences. And the list goes on!

To put the importance and uniqueness of thermodynamics into perspective we quote two outstanding scientists.

1.1 The verdict on thermodynamics

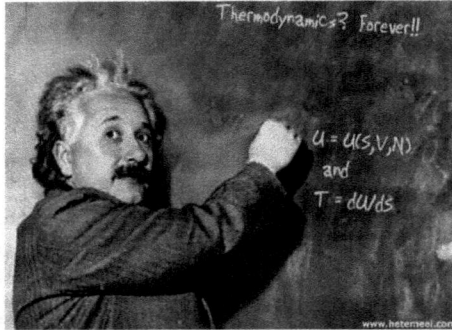

Albert Einstein

"A theory is the more impressive the greater the simplicity of its premises, the more varied the kinds of things that it relates and the more extended the area of its applicability. Therefore classical thermodynamics has made a deep impression on me. It is the only physical theory of universal content which I am convinced, within the areas of the applicability of its basic concepts, will never be overthrown."

Sir Arthur Stanley Eddington

"The law that entropy always increases – the second law of thermodynamics – holds I think, the supreme position among the laws of Nature. If someone points out to you that your pet theory of the universe is in disagreement with Maxwell's equations - then so much worse for Maxwell equations. If it is found to be contradicted by observation - well these experimentalists do bungle things sometimes.

But if your theory is found to be against the second law of Thermodynamics, I can give you no hope; there is nothing for it but to collapse in deepest humiliation."

1.2 The need for a macroscopic description

To understand that a macroscopic description of matter is actually meaningful and needed, let us look at two simple examples.

A solid: The atoms or ions making up a solid are confined to regular lattice sites by the sum total of the interactions with all their neighbors. They are however not at rest at the bottom of their respective potential wells, but vibrate around these minima incessantly. To estimate these vibrational frequencies, note that the potential wells must have widths of the order of the interatomic spacing, a, in a solid, i.e. of the order of a few Å. On the other hand, the depth of these potentials is the energy needed to remove an atom from the solid, i.e. the cohesive energy, V_c, of the solid which is typically a few electron volt ($1 \text{ eV} = 1.602 \times 10^{-19}$ J), e.g. 6.2 eV for tungsten. If we approximate the site potential by a harmonic oscillator

$$V(r) = \frac{1}{2}kr^2 \qquad (1.1)$$

we get

$$k = 2V_c/a^2 \qquad (1.2)$$

and the frequency of vibration

$$\nu = \sqrt{k/m}/2\pi \qquad (1.3)$$

where m is the mass of one solid atom, say tungsten. Putting in numbers we get $\nu \approx 10^{13}$ s^{-1} as a typical vibrational frequency in the solid.

Now, the energy stored in these vibrations is proportional to the square of their amplitudes. Although every 10^{-13} s the energy is redistributed among those 10^{23} atoms in the solid, the total energy is constant, as long as the solid is isolated. If we heat up the solid, the vibrations increase in amplitude and the total energy goes up. Not only that, the volume of the solid increases; this is called thermal expansion. Although on a microscopic scale, the state of the system, i.e. the position and velocities of the constituent atoms, changes every 10^{-13} s, we can still meaningfully describe the thermodynamic state of the system by specifying, so far, three quantities, namely its **mass, volume**, and **energy**. Not only that, it would be quite impossible and utterly useless to follow the motion of all 10^{23} atoms anyway.

The ideal gas: Let us look, as another example, at a gas at room temperature and atmospheric pressure enclosed in a volume. Its density is about 10^{25} molecules per m^3. A typical velocity of a gas molecule is the speed of sound in the gas, i.e. about 10^3 m/s. It will collide every 10^{-10} s with another molecule in the gas resulting in a redistribution of momentum and energy. Yet, the total energy is conserved again and is a measure of the macroscopic state of the gas.

1.2.1   *Ideal gas: A macroscopic description*

A gas of mass M is confined in a container of volume V. It exerts a pressure P - force per area - on the walls of the container. If the density of the gas, i.e. mass per volume, is "low" enough it was found by Boyle that

$$P \sim \frac{M}{V} \tag{1.4}$$

provided the temperature is held constant. Boyle's experiment is simple: he took a flask with a movable piston and filled it with a gas which he then immersed into a large bucket of water of constant temperature. By putting various weights on the piston, Boyle was able to change the pressure and measure the corresponding change in volume. He found that volume changes inversely to the pressure. Doing a similar experiment, Charles and Gay-Lussac found that at constant pressure the volume of the flask grew as its temperature increased. Combining these two findings and expressing the mass of the gas in terms of the number of moles N, i.e. $M \sim N$, results in the "ideal gas law"

$$PV = NRT \tag{1.5}$$

It was found that the constant of proportionality, R, in this equation is the same for all gases at low density. This is a remarkable finding: no matter what the gas is made of, be it oxygen, gasoline vapor, or a high temperature vapor of several metals all mixed together, the relation between pressure, density and temperature is the same.

Remark 1.1. "Ideal Gas" is actually a misnomer as the ideal gas law applies to real gases such as the air in a lecture hall. What made the term "ideal" stick is the fact that it applies to all gases at low densities irrespective of their composition. To understand this remarkable fact, recall that a gas consists of many particles that move around at high speeds summing up to a considerable kinetic energy of all the particles. In addition to the kinetic energy we must also consider the potential energy of the interactions between the particles because only the sum of kinetic and potential energy is conserved according to Newton. However, the collisions are rare and thus the interactions are small and the potential energy can be neglected. Thus in an ideal gas the interaction energy is negligible compared to the kinetic energy. And because the atom- or molecule-specific characteristics of a gas are expressed in their interactions (two helium atoms interact differently from two oxygen molecules), the fact that at low density the potential energy of interactions is negligible makes it "ideal".

Remark 1.2. The ideal gas law, although so simple, already demonstrates an inherent usefulness of thermodynamic relations. It allows us to calculate further properties of a gas, e.g.

(a) its isothermal compressibility, the relative change in volume when changing pressure at fixed temperature,

$$\kappa_T = -\frac{1}{V}\frac{\partial V}{\partial P}|_T = \frac{1}{P} \qquad (1.6)$$

(b) or its thermal expansion coefficient, the relative volume change when changing temperature at constant pressure,

$$\alpha = \frac{1}{V}\frac{\partial V}{\partial T}|_P = \frac{1}{T} \qquad (1.7)$$

Both quantities are relevant when designing devices involving expanding gases.

The variables pressure P, volume V, and mole number N (or mass) are known from mechanics and can be measured by mechanical means. We have introduced temperature without proper definition and thus without a prescription on how to measure it. Temperature is understood to provide a measure for our sense of hot and cold. As well, two objects in contact with each other for a long time are said to have the same temperature and to be in equilibrium with each other.

1.2.2 *Measurement of temperature*

To design a measuring device for temperature we need to identify material properties that depend on it. Examples are: (i) the expansion of gases, liquids and solids with temperature, (ii) changes in conduction of metals and semiconductors with temperature, (iii) changes in color with temperature, and many more as listed in the table:

Thermometer	Thermometric property
Gas (constant volume)	Pressure
Electric resistor (constant tension)	electric resistance
Thermocouple (constant tension)	Thermal EMF
Paramagnetic salt	magnetic susceptibility
Blackbody radiation	radiant emittance

1.2.2.1 *Constant volume gas thermometer*

The simplest thermometric property is no doubt the change in pressure in a gas at constant volume with temperature; it is also the most fundamental because it applies to all gases at low density. This is the basis of the Constant Volume Gas thermometer depicted below.

It consists of a bulb (originally glass, but these days usually platinum or a platinum alloy) connected to a U-tube which itself is connected to a flexible tube leading to a reservoir with a liquid such as mercury or alcohol[1]. The reservoir is open to ensure that the reference pressure is the atmosphere, or an artificial atmosphere of an inert gas to avoid contamination of the liquid. The volume of the gas is kept constant by adjusting the height of the mercury column M to touch a mark on the tube called the indicial point. The difference h between the levels in the two columns of thee U-tube determines the pressure in the bulb: $P = P_{atm} + \rho g h$ where P_{atm} is the atmospheric pressure, ρ is the mass density of the mercury, and g is gravity. To calibrate the thermometer we must find a reference

---

[1]The kind of liquid used depends on the temperature range of interest. For example, water would not do below freezing or above boiling.

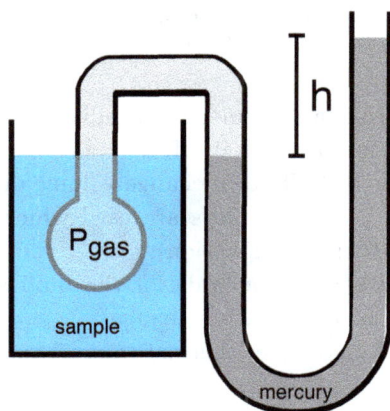

Fig. 1.1 Constant volume gas thermometer. Pressure relative to the atmosphere is measured based on the height of the liquid column. This pressure is related to the gas temperature by the ideal gas law.

or standard system to which we assign a numerical value for its temperature. Any reference system would do as long as it is easily reproducible and universally accessible. In 1954 it was decided worldwide that the standard reference point is the triple point of water, i.e. the state of ultra pure water with a well-defined isotope composition in which ice, liquid, and vapor co-exist. At the triple point ice cubes float in liquid water with saturated water vapor above it; this state will remain the same as long as it is undisturbed: no melting, no freezing, no evaporation. This happens at a partial water pressure of $P_t = 611.73$ Pa and also at a unique temperature to which the value $T_t = 273.16$ K is arbitrarily assigned. The symbol K has been adopted to honor Lord Kelvin, a great thermodynamician who perfected the gas thermometer.

Bringing the gas bulb of the thermometer in contact with water at its triple point we measure the pressure in the bulb, and call this pressure P_t. Next we bring the thermometer into contact with the system whose temperature we want to determine, measure again the pressure in the gas bulb, call it P, and calculate its temperature according to the ideal gas law

$$T = \frac{T_t}{P_t} P \qquad (1.8)$$

A series of such measurements is performed at smaller and smaller gas densities in the bulb to extrapolate to the limit of vanishing density. Marks on the U-tube will indicate the temperature scale.

To achieve maximum accuracy a number of corrections must be made such as:

(1) The gas in the U-tube and connecting capillary is not necessarily at the same temperature as the bulb itself;
(2) The capillary and the bulb itself change volume due to thermal expansion (that is why platinum alloys are used which can be designed to undergo minimal thermal expansion);
(3) Some gas is adsorbed to the walls of the bulb and the capillary in different amounts as temperature changes;
(4) The mercury itself undergoes thermal expansion.

Sophisticated Constant Volume Gas thermometers are maintained at national bureaus of standards such as NIST in Washington DC, NRC in Ottawa and are used to calibrate household and industrial thermometers. Practical temperature readings are made in degrees Celsius ($0°$ C = 273.15 K). $100°$ C is the temperature at which pure water boils and evaporates at standard atmospheric pressure.

Having properly defined temperature we can now also assign a (measured) value to the universal gas constant per mole

$$R = 8.31441 \text{ JK}^{-1}/\text{mol} \tag{1.9}$$

Remark 1.3. At the triple point of any gas-liquid-solid system an arbitrarily small decrease in temperature or increase in pressure will cause all the material to freeze. Look at the phase diagram in Figure 1.2 to follow this argument. A triple point cell is made by adding the right amount of material so that at the triple point temperature it leads to the triple point pressure. This is tricky, but can be done for many liquids to fit various temperature regimes.

Substance	Acetylene	Argon	graphite	CO_2	CO	Ethanol
$T_t[K]$	192.4	83.81	4766	216.55	68.10	150
$P_t[Pa]$	120	68.9	10132	517	15.37	4.3×10^{-7}

Substance	hydrogen	mercury	nitrogen	platinum	water
$T_t[K]$	13.84	234.2	63.18	2045	273.16
$P_t[Pa]$	7.04	1.65×10^{-7}	12.6	2.0×10^{-4}	0.6117

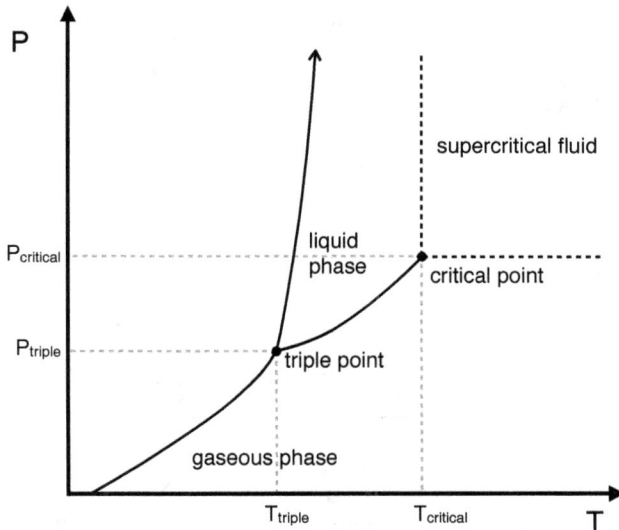

Fig. 1.2   A typical single component equilibrium phase diagram. For points away from the transition line, the system is homogenous. For those on the transition line, coexistence of two phases such as gas-liquid and liquid-solid is possible. At the conditions corresponding to the triple point, all three phases (gas, liquid, and solid) can exist simultaneously.

1.2.2.2   *Blackbody radiation*

Any object (in equilibrium at some temperature) will emit electromagnetic radiation with a characteristic power spectrum called blackbody radiation. A small hole in a cardboard box of any color appears black at room temperature, a clay pot in a kiln glows red, and steel being forged may be red to white. The explanation of this spectrum lead Max Planck in 1900 to the discovery of quantum physics. He derived his celebrated equation for the spectral radiance, the energy per unit time and unit surface area in a wavelength interval between λ and $\lambda + d\lambda$,

$$I(\lambda, T) = \frac{2hc^2}{\lambda^5} \frac{1}{\exp[hc/(\lambda k_B T)] + 1} \tag{1.10}$$

where h is Planck's constant and c is the speed of light. Note that this spectrum only depends on temperature!

Thus measuring an emission spectrum and fitting it to Planck's formula determines the temperature of the body. No contact with the object is needed; the measurement is remote and does not disturb the object. This

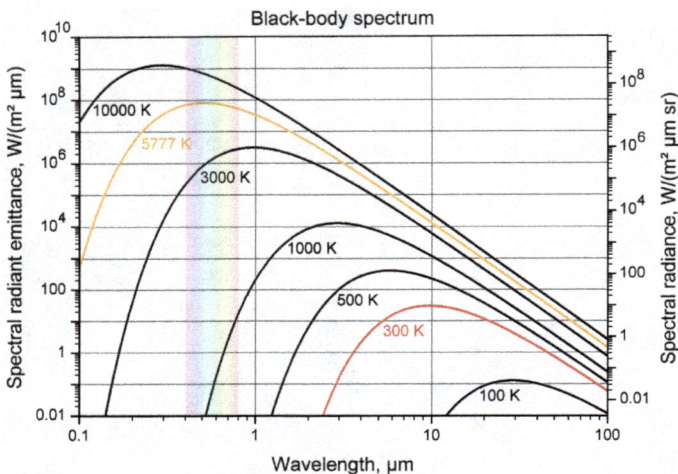

Fig. 1.3   The spectral radiance (or emittance) of a blackbody as a function of wavelength according to Planck's law.   For a given temperature, a blackbody radiator will emit radiation across a continuum of wavelengths. Changes in temperature result in both an increase in the total output as well as a shift in the relative contribution from different wavelengths. Image adapted from [WikimediaCommons (2006)].

is the way the extreme temperatures in a coke oven, a nuclear blast, or on the surface of the Sun can be measured. Furthermore, it was the way that Penzias and Wilson detected the cosmic background radiation of the universe that had been predicted as a remnant of the Big Bang - the latest measurements yield a temperature of 2.725 K, rather cold but largely isotropic in the universe.

The spectrum from the sun is distorted due to absorption of some of the light on its way to earth; but overall a fit with Planck's law yields a good estimate of the temperature of the Sun's surface of 5323 K.

Remark 1.4. Why is the hole in a box at room temperature black? First of all, the color of the box itself is determined by those frequencies from the light that fall onto it that are not absorbed. Any light that hits the hole will be reflected off or absorbed by the interior side of the walls. Thus it thermalizes with the walls before it can escape from the hole. What is the characteristic wavelength or frequency of light at the maximum of the Planck distribution? Differentiation leads to $\lambda_{\max} = 2.89776829\frac{1}{T} \times 10^6$ nm. Thus at room temperature we have $\lambda_{\max} \approx 10^4$ nm $= 0.01$ mm; this is a wavelength which our eyes do not register. On the other hand for sunlight

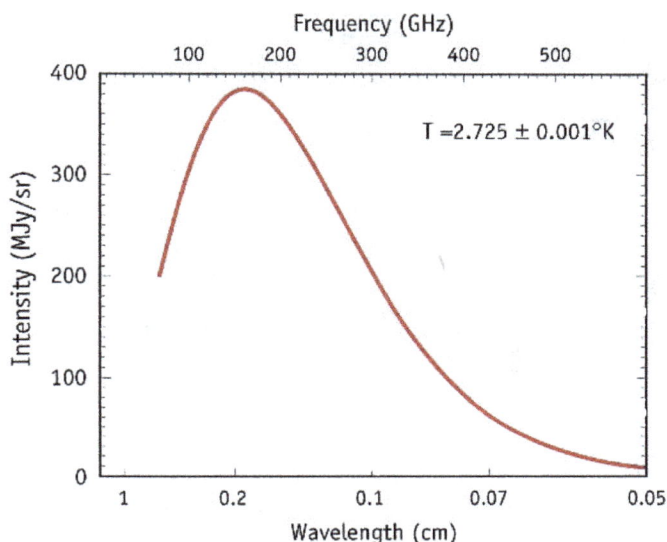

Fig. 1.4   Telescopes detect low levels of radiation from regions of space that are seemingly empty. The wavelength dependence of this radiation is consistent with a cosmic background temperature of roughly 3 K. Note that this curve is experimental data, not simply a curve fit.

at $T_{sun} = 5323$ K we get at the maximum of the emission spectrum $\lambda_{max} \approx$ 500 nm which we register as green light. As our eyes have adapted to sunlight over eons of evolution green light is the easiest and thus most pleasant light whereas for red light the eyes are less adapted. Therefore red signals danger in traffic and green is a go-ahead.

1.2.3   *Ideal gas: A microscopic description*

Thermodynamics is a macroscopic theory and thus does not require input from the microscopic dynamics of the molecular or atomic building blocks making up the macroscopic system. Still, it is useful at this stage to have a microscopic view of an ideal gas to contrast it with the macroscopic view just presented. We can do this in the framework of some simple kinetic gas theory or statistical mechanics of a dilute gas. These ideas go back to Daniel Bernoulli (1738), rejected at the time but resurrected in the early 1800's in the form presented here.

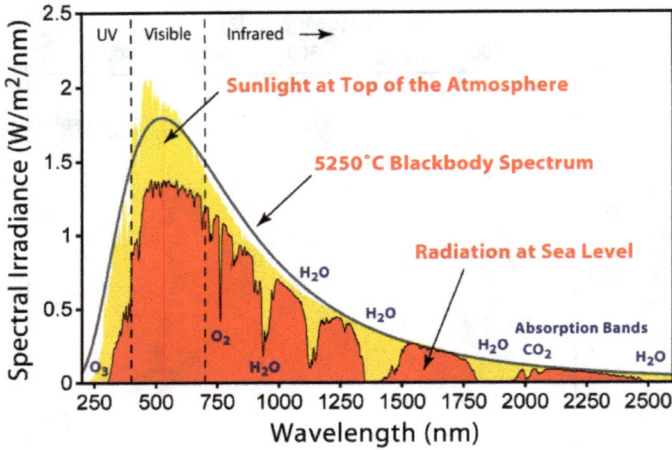

Fig. 1.5   The temperature of the surface of the Sun can be inferred from the radiation it emits (a fit of 5323 K is shown). At the surface of the Earth, this spectrum is modified from what is measured in space due to the presence of the atmosphere. Different molecular species selectively absorb energy, resulting in gaps in the spectrum. The Sun emits the most power at wavelengths corresponding to what humans refer to as visible light. This is an evolutionary artifact - human eyes have evolved to be sensitive to the light emitted by our Sun. Image adapted from [WikimediaCommons (2007a)].

Let us consider a dilute gas of $\widetilde{N}$ atoms, each of mass m, contained in a cubic volume $V = L^3$. The ith atom, treated as a mass point for simplicity, has, at a given time, a velocity $\mathbf{v}^{(i)}$ with a component $v_z^{(i)}$ in the z-direction perpendicular to one of the walls of the cube.

When it collides with the wall, it is reflected and moves away from it with the z-component of its velocity reversed, so that its total momentum is changed by $-2m \left| v_z^{(i)} \right|$ imparting an impulse $2m \left| v_z^{(i)} \right|$ to the wall (Newton's second law). The force on the wall is given by the product of the momentum change per atom times the number of impacts this atom makes with the wall per unit time. The latter is given by $\left| v_z^{(i)} \right| / 2L$. Thus the pressure on the wall (force per unit area) is

$$P = \sum_{i=1}^{\widetilde{N}} 2m |v_z^{(i)}| \frac{|v_z^{(i)}|}{2L} \frac{1}{L^2} = \sum_{i=1}^{\widetilde{N}} \frac{m(v_z^{(i)})^2}{L^3} \qquad (1.11)$$

or

$$PV = \sum_{i=1}^{\widetilde{N}} m(v_z^{(i)})^2 \qquad (1.12)$$

To evaluate the sum exactly we would need the velocities of all $\widetilde{N}$ atoms, typically 10^{19} per cm^3. This is impossible, in particular if you consider that in a gas the atoms collide with each other on average every nanosecond, changing their velocities each time. It is not only impossible to know this in practice but also unnecessary. All we need is the most probable distribution of such velocities; this function can be measured in a molecular beam experiment. For our purposes here, even that much information is not needed. We can introduce a mean squared velocity of the $\widetilde{N}$ atoms

$$\overline{\mathbf{v}^2} = \frac{1}{\widetilde{N}} \sum_{i=1}^{\widetilde{N}} \left[(v_x^{(i)})^2 + (v_y^{(i)})^2 + (v_z^{(i)})^2 \right] \tag{1.13}$$

If the velocity distribution of the atoms is isotropic in space (which it is if there is no external potential such as gravity around, or is negligible) then the mean square averages of the three velocity components are equal and we get

$$PV = \frac{2}{3} \left(\frac{1}{2} M \overline{\mathbf{v}^2} \right) \tag{1.14}$$

$$= \frac{2}{3} U$$

U is the total kinetic energy of the atoms flying around in the gas; it is called in thermodynamics the **internal energy** because it accounts for the motion of the constituent particles of our macroscopic system (the gas in V). Also note that with m, the mass of one of the $\widetilde{N}$ gas particles, the total mass of the gas is $M = \widetilde{N}m$.

Remark 1.5. If the interaction between gas particles is no longer negligible as their density increases, the internal energy would also have a contribution from the potential energy of that interaction. But in any case the internal energy, say of an ice cube, has nothing to do with its kinetic energy as a whole as it is somehow hurled through space.

Comparing this to the ideal gas law we see that the internal energy is related to temperature

$$U = \frac{3}{2} NRT$$

$$= \frac{3}{2} \widetilde{N} k_B T \tag{1.15}$$

Here

$$k_B = \frac{R}{N_{\text{Avogadro}}} = 1.38066 \times 10^{-23} \text{ J/K}$$

$$N_{\text{Avogadro}} = 6.02205 \times 10^{23} / \text{mol} \tag{1.16}$$

k_B is called Boltzmann's constant; it is the universal gas constant per atom or molecule. Remember that the mole number is the total number of particles in the system divided by Avogadro's number (which is the number of particles in one mole).

This derivation makes clear that to describe a macroscopic system we do need additional variables, such as temperature, which account for the effect of the myriads of degrees of freedom of the atoms or molecules which we choose to ignore in a macroscopic or thermodynamic approach.

Example 1.1. What is the thermal speed of a helium atom at room temperature?

Combining (1.14) and (1.15) we get for the root mean square (rms) speed in one direction at $T = 300$ K with $m = 4 \times 1.67 \times 10^{-27}$kg the mass of a helium atom

$$\sqrt{\overline{\mathbf{v}^2}} = \sqrt{3k_BT/m} = 1,367 \text{ m/s} \tag{1.17}$$

This compares well with the speed of sound for helium $v_{sound} = \sqrt{\gamma k_BT/m} = 927$ m/s. Here γ is called the adiabatic constant and has a value of $5/3$ for an atomic gas.

Does helium escape into outer space from the Earth's atmosphere?

The escape velocity from a gravitational body is obtained by equating the initial upward kinetic energy to the gravitational energy at a distance R from the center of the Earth: $(m/2)v_{esc}^2 = GmM/R$ or $v_{esc} = \sqrt{2GMR}$. The Earth's mass is $M = 5.97 \times 10^{24}$ kg, so an object can escape if its speed at the Earth's surface ($R = 6400$ km) is $v_{esc} = 11.2$ km/s, much larger than the average speed of helium. Yet, helium escapes anyway. There are several contributing factors of which we just list two: (1) the speed of any gas particle varies from zero to many times the average speed. Thus energetic atoms can have speeds larger than the escape velocity. (2) In the ionosphere and exosphere at heights above 300 km the temperature is in excess of 1200 K so the escape is possible. In addition, the solar wind sweeps away light atoms such as helium.

Example 1.2. How often do helium atoms collide with each other at room temperature and atmospheric pressure and how far do they travel between collisions?

The flux of particles in a gas in any direction is given by the particle density and the average velocity: $j_{th} = (\widetilde{N}/V)\sqrt{\overline{\mathbf{v}^2}} \approx 2.7 \times 10^{25} \times 10^3$ particles per square meter and second. Approximating a gas particle by a sphere of radius r those particles that are within a disc of area πr^2 will collide. With $r \approx 1$ $\mathring{A}$ we thus get for the number of collisions per second $\tau_{\mathrm{coll}}^{-1} \approx 2 \times 10^9$ s^{-1} or $\tau_{\mathrm{coll}} \approx 5 \times 10^{-10}$ s. The distance traveled between any two collisions is called the mean free path; it is $\lambda_{\mathrm{mfp}} = \tau_{\mathrm{coll}} \sqrt{\overline{\mathbf{v}^2}} \approx 1$ μm.

1.3 What will thermodynamics do for you?

Given a macroscopic system you can do a lot of experiments on it: you can measure the length of a steel beam as you heat it up, keeping the pressure constant thus obtaining the expansion coefficient; this is important information when an engineer designs a bridge because its steel structure will expand in summer and it must still fit into its foundation. Similarly you could measure the length as pressure is varied, keeping the temperature constant; this is important knowledge for constructing high pressure or vacuum vessels. You could also do either experiment not keeping pressure or, respectively, temperature constant. The question arises how many such experiments do you have to perform to know everything about the macroscopic or thermodynamic properties of a system. Infinitely many? Thermodynamics will tell you that there are very few such independent experiments, e.g. for a simple system just three. All other data you can generate from these very few experiments by using thermodynamic formulae. Thus thermodynamics facilitates an enormous reduction in work. Not only that, by providing formulae to calculate the results of related experimental set ups, it has predictive power, e.g. in the construction of engines. Lastly, if you were to perform redundant experiments, you could find out whether you are making systematic errors, as the thermodynamic formulae provide consistency checks. Many new physical phenomena have been found in this way, e.g. Planck's law of blackbody radiation and thus the discovery of quantum physics. As Einstein said, all other physical theories must conform to thermodynamics.

1.4 What does thermodynamics not do for you?

As we said, thermodynamics ignores all those myriads of degrees of freedom associated with the constituent atoms and molecules of a large system, and deals with just a few macroscopic variables. In thermodynamics we take the fact that lead is more pliable than steel or that you can liquify carbon dioxide as a given and use it to make predictions for other properties of lead, steel, and carbon dioxide. To understand why this is so we need to incorporate the electronic properties of lead, steel, carbon dioxide etc. into a theory. This is done in statistical mechanics where one actually starts from Newton's or Schrödinger equations for all degrees of freedom, and then performs a statistical average over them to arrive at thermodynamics. In the course of doing this, one also gets, in favorable model cases such as a dilute gas or a harmonic solid, expressions for the characteristic properties such as expansion coefficients, compressibilities, etc. Thus statistical mechanics provides the link between the microscopic and the macroscopic world. The reason why you do not learn about statistical mechanics right away, is that (i) you must first know what to derive, and more importantly (ii) in statistical mechanics you solve simplified models of systems, whereas thermodynamics provides a theoretical framework to which all systems must adhere.

1.5   Problems

Problem 1.1. *A gas is confined to a cylinder of volume $V = 100$ l at a pressure $P = 2$ MPa.*

(a) Calculate the rms velocity of the gas particles.
(b) Assume the gas is molecular nitrogen. How many moles are in the gas cylinder? At what temperature is the gas cylinder?

Problem 1.2. *In nuclear fusion as it occurs in stars, fusion bombs and hopefully soon in fusion reactors, temperatures must be of the order of 10^9 K before deuterons (the nuclei of deuterium consisting of a proton and a neutron) smashing into each other have a chance to fuse into helium. What is the rms speed of deuterons at that temperature. How does that compare to the speed of light?*

Problem 1.3. *It is estimated that the universe contains a total mass of 10^{80} proton masses ($m_p = 1.67 \times 10^{-27}$ kg). Its present volume is estimated to be about 10^{80} m^3. Its temperature is that of the cosmic background radiation, $T = 2.7$ K.*

(a) Assume that the mass is distributed uniformly ignoring aggregation into stars and galaxies. What is the matter density of the universe and what is the average distance between any two protons?

(b) Because of the low density we can treat the protons as an ideal gas. What is the pressure?

Problem 1.4. A hot air balloon carries a payload of mass M_{load} (including the skin of the balloon and the gondola). The hot air inside the balloon is at a temperature T_b and the temperature of the surrounding atmosphere is T_a. The balloon will expand until the pressures inside and out are equal. The balloon will remain stationary once the buoyant force on the balloon, given by the weight of the replaced cool air, is compensated by the weight of the hot air and of the payload.

(a) Treating the air as an ideal gas and the balloon as a sphere, derive a formula for the diameter D of the stationary balloon.
(b) Take $T_a = 20°$ C, $T_b = 40°$ C and the molar mass of air as $M_{air} = 1.2$ kg/mol. What is D for a payload of $M_{load} = 3000$ kg?

Problem 1.5. A gas cylinder is divided into two volumes V_1 and V_2, separated by a thin membrane. The first chamber contains helium at a temperature T_1 and pressure P_1; the second chamber is initially evacuated. The membrane is punctured to allow the gas to also fill the second chamber. Because nothing is done to the system (aside from puncturing the membrane) the internal energy before and after the puncture remains the same; this is called a free expansion.

Treating the gas as ideal with an internal energy $U = (3/2)NRT = (3/2)PV$, determine the final temperature T_f, volume V_f and pressure P_f.

Problem 1.6. In the gas cylinder of Problem 1.5 the volume V_1 contains helium at pressure P_1 and temperature T_1. Also, volume V_2 contains argon at pressure P_2 and temperature T_2. Puncturing the membrane will now allow the two gases to mix.

Determine the final temperature T_f, volume V_f and pressure P_f.

Problem 1.7. On a hot summer day $(T = 30°$ C) car tires are inflated to their normal pressure of 32 psi. As the temperature drops so does the tire pressure.

(a) At what temperature has the pressure dropped to 30 psi?
(b) What will it be at $T = -10°$ C?

Problem 1.8. The vibrational frequency of an atom around its equilibrium lattice position in a crystalline solid is $\nu = \sqrt{k/m}/2\pi$. Here $k = 2V_c/a^2$

is the force constant of the (approximately) harmonic restoring force. Aluminum has a mole number 27. What is the mass of one aluminum atom?

(a) For a typical vibrational frequency $\nu = 5 \times 10^{12}$ s^{-1} what is k?
(b) Solid aluminum has a density of 2.7 g/cm^3. What is the average distance between atoms?
(c) What is your estimate for the cohesive energy V_c of aluminum in Joules and in electron volts?

Problem 1.9. *The resistance of a doped germanium crystal is found to be well represented by the equation* $\ln R = 10.8031 - 9.0291 \ln T$ *where, for this calibration, T was measured by a gas thermometer.*

(a) In a bath of liquid helium the resistance is measured to be $R = 218$ Ω. What was the temperature?
(b) Make a log-log plot (not ln-ln) for the range 200 Ω to 30,000 Ω.

Chapter Summary

(1) The Verdict on Thermodynamics by Einstein and Eddington!
(2) Looking at the ideal gas from a macroscopic and microscopic point of view.
(3) Measurement of temperature by the ideal gas thermometer and blackbody radiation.
(4) What can thermodynamics do for you?

Chapter 2

The Structure of Thermodynamics

Thermodynamics is more than a theory. It is a general way of thinking about systems, of tying together and ordering facts and properties, and of establishing laws that all systems must satisfy under certain conditions. It is a minimalist's approach: what is the minimum set of data about a system needed to understand and predict other properties?

We introduce the notion of degrees of freedom to count the number of independent ways a particle can move. A point particle has three degrees of freedom in three-dimensional space. From our discussion so far we can then conclude that of the 10^{23} or so degrees of freedom of a macroscopic piece of matter, only a few, such as energy and volume, are necessary to describe its macroscopic state. However, the myriads of other degrees of freedom are not irrelevant. Look, as an example, at the mechanical work that we do by pushing a piston into a cylinder. As a result the gas will have a higher energy, i.e. all the molecules in it will move about with a higher average kinetic energy. We could also increase the energy in the system by heating the cylinder (with a fixed piston) with a torch. In that situation the energy transfer to the myriads of "irrelevant" degrees of freedom is called heat transfer. One could say that thermodynamics is that part of physics concerned with processes in large systems involving heat transfer.

Thermodynamics deals with large systems in and close to equilibrium.

2.1 Large systems

All physical systems consist of many atoms or molecules; the stock market consists of many stocks and investors. To treat a system such as the still air in a classroom as a large thermodynamic system, we need to know that the density of air in one corner of the room is the same as in another. So imagine that we put some imaginary walls around a liter $(= 10^3 \text{ cm}^3)$ of air in one corner. If that parcel of air is in equilibrium it will contain $\mathcal{N} = 2.687774 \times 10^{22}$ particles according to the ideal gas law. How accurately can we establish this number, and how different is that number for another parcel of air of the same volume and at the same temperature and pressure somewhere else in the room? Elementary statistics tells us that in a random sampling (counting) the relative error is $\delta\mathcal{N}/\mathcal{N} = \mathcal{N}^{-1/2}$ or 10^{-11}. This is the likely error we would make statistically if we were foolish enough to try to count. For a physical system this is called relative fluctuations. Thermodynamics deals with systems large enough that the relative fluctuations can be neglected. Note that on this argument thermodynamics can deal with any object that is large enough to see.

2.2   Macroscopic variables

A large system consists of myriads of particles that move around incessantly, freely in a gas with occasional collisions, and vibrating around their lattice positions in a crystalline solid. Despite the constant molecular changes that take place, there are certain variables and properties that are time-independent in an isolated system. Isolation means that the walls around the system, the cylinder containing a gas or a liquid, are fixed and do not allow the exchange of mass or energy.

> **An isolated system is characterized by a small set of macroscopic or thermodynamic variables such as the volume V, the total mass, measured via mole numbers N_i for different chemical components, and the total energy U. Total electric charge, electric and magnetic dipole moments and others are further thermodynamic variables. Note that these variables are already familiar from classical physics.**

2.2.1   *Equilibrium: A question of time and history*

Let us take a macroscopic system and isolate it completely from the rest of the world by enclosing it in rigid walls that are also thermally insulating, e.g. a gas in a steel cylinder surrounded by a styrofoam box. This assigns particular values to the thermodynamic variables V, N, and U. After the initial perturbation of preparation has died down the system will settle in a macroscopically time-independent state called **equilibrium**. Of course the atoms in the system will still undergo their microscopic motion colliding with each other and exchanging energy, but the overall energy is conserved after you have isolated the system.

Example 2.1. Ideal Gas: The average speed of a gas particle is roughly the speed of sound (the only characteristic speed in a gas!) which in air is $c_s = 331$ m/s. On average an air particle at room temperature will collide with another one every nanosecond, this is called the collision time τ. The initial perturbation must die out throughout the volume V; this is achieved after sound waves have traversed the volume a few times which takes roughly $V^{1/3}/c_s$. After that time no further macroscopic evolution is observed and the gas is in macroscopic equilibrium although each gas particle still travels around randomly.

Thus the **equilibrium** state of any system is by definition **time independent**. But that is not enough! It must also be **independent of its past history** otherwise it would not be unique. Not all systems that seem to be in equilibrium actually are. Without care one can get fooled easily.

Example 2.2. Graphite and Diamonds: The equilibrium state of carbon below about 4000 K and 1 GPa is graphite, i.e. a crystalline state in which carbon atoms are arranged in a hexagonal pattern in planar sheets which are separated by 3.9 Å. This atomistically "large" separation is the reason why carbon sheets can be cleaved easily. However, if you expose graphite to high temperatures and high pressures ($P > 10$ GPa) the atoms re-arrange themselves in a diamond structure. This happens in Nature at magma inclusions deep in the earth's crust and leads to diamond deposits. It can also be done in the laboratory producing synthetic diamonds. Diamond is the equilibrium form of carbon at these high temperatures and pressure but it is NOT in equilibrium at standard temperature and pressure (STP: room temperature $T = 20^0$ C $= 293.15$ K, and a pressure $P = 1$ atm $= 101.325$ kPa)! To prove this, heat a diamond to a temperature of 5000 K to vaporize it and then cool it down fast: it transforms into a black powder which again is NOT an equilibrium state.

Conclusion: the equilibrium state is unique for every system. There may be long-lived metastable states but those are not at equilibrium.

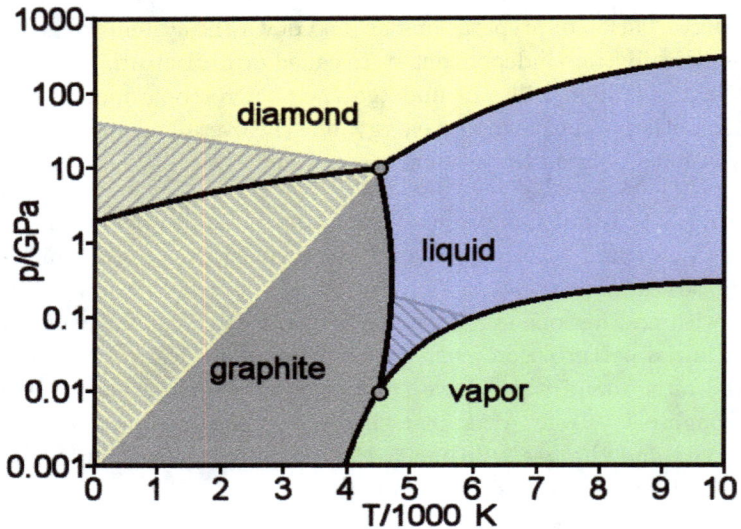

Fig. 2.1   A single component, elemental phase diagram of carbon. This diagram denotes the equilibrium phases of carbon over a wide range of temperatures and pressures. Note that at ambient conditions, diamond is not a stable structure in the thermodynamic sense. Image adapted from [WikimediaCommons (2005)].

2.2.2   *The fundamental relation: Entropy*

There are infinitely many macroscopic systems that you can prepare in such a way that they have the same values of V, N_i, and U but are physically distinct. To characterize a given system and distinguish it from all others we require knowledge of at least one further property, a function of these variables, unique for each system. A clue comes from the fact that in a macroscopic approach we completely ignore the internal dynamics of the myriads of atoms and molecules that make up a system. It turns out that what is needed is just one more variable called the entropy, S. It is a measure of the internal motion in a system, *i.e.* of its relative state of disorder (a gas of atoms) or order (a crystalline solid of the same material) or anything in between. To clarify this point further recall that in a gas the particles can be anywhere in the enclosing container. On the other hand in a crystalline solid these same particles are arranged on fixed lattice sites about which they oscillate with small amplitudes. In the crystal we know more about where the particles actually are; the crystal is more ordered than a gas. In thermodynamics the entropy (derived from the greek $\epsilon\nu$-

"in" + $\tau\rho o\pi\eta$ "turning") was introduced by Rudolf Clausius in 1862 as a measure of disorder. He used ice melting as a example of, in his words, "entropy increasing due to the disintegration of the molecules of the body of ice". In information theory, entropy or Shannon entropy is a measure of the uncertainty associated with a random variable, introduced by Claude E. Shannon in his 1948 paper "A Mathematical Theory of Communication". The complementary measure for order is called negentropy and is also used in information theory.

> **The entropy is a function of all thermodynamic variables, $S = S(U, V, N_1, N_2, ..., N_r)$ for a system consisting of r different components. This function, called the Fundamental Relation in the entropy representation, is unique for a given system and defines all its equilibrium states uniquely.**

Macroscopic properties of physical systems vary continuously when control parameters are changed. As an example, lowering the temperature of water makes it colder but it remains liquid down to 0° C when it changes its physical properties abruptly and discontinuously by freezing into solid ice. As a precaution we should therefore demand of the **entropy** function that it is **piecewise continuous and differentiable**. We will learn in Sections 4.2 and 4.3 that S must be a monotonically increasing function of U. In addition, in Section 3.6 we will state that S must approach zero as the temperature goes to zero.

Remark 2.1. The thermodynamic variables that we have introduced so far are all **additive** or **extensive**: consider a system that consists of two isolated subsystems

$V^{(1)}$
$U^{(1)}$
$N_i^{(1)}$
$S^{(1)}$

$+$

$V^{(2)}$
$U^{(2)}$
$N_i^{(2)}$
$S^{(2)}$

$\Longrightarrow$

$V = V^{(1)} + V^{(2)}$
$U = U^{(1)} + U^{(2)}$
$N_i = N_i^{(1)} + N_i^{(2)}$
$S = S^{(1)} + S^{(2)}$

with $S^{(k)} = S^{(k)}(U^{(k)}, V^{(k)}, N_1^{(k)}, N_2^{(k)}, ..., N_r^{(k)})$, then the total system has $S = S(U, V, N_1, N_2, ...) = S^{(1)} + S^{(2)}$. Another way of looking at it is to double the volume of a system keeping its thermodynamic properties the same. Then one must obviously double all the other extensive variables as well.

2.3   Measurement and walls

2.3.1   *Walls*

Every macroscopic system is finite and thus has boundaries or walls that separate it from its surroundings. To change the state of a system, one has to manipulate the walls. Thus we need a classification of walls according to the manipulation they allow.

We call a wall **rigid** if it does not allow a volume change. To change the volume V, the wall must be **moveable**. Note that in order to assign a volume to a system in the first place, there must exist walls for the system that are rigid for the duration of the volume measurement.

To assign a mole number N_k for species k, there must be walls that are **impermeable** to component k. To exchange material with the surroundings; the walls must be **permeable** to component k. For a gas in a container a permeable wall is typically a valve that connects the system to a gas supply cylinder, but it could also be a membrane permeable for a given component. This is to a large measure the basis for life: cell membranes allow certain ions to permeate and others not.

To assign a value to the internal energy U, one must have walls restrictive to energy transfer. We note that rigid walls are restrictive to mechanical work such as the work done by compression or by stirring a liquid or gas (the surface of a spoon or of a paddle wheel used to stir a liquid is part of the surface surrounding the system). In addition to changing the internal energy by work (i.e. mechanical energy transfer), we can also change it by heat transfer (putting a candle under it). Walls restrictive to heat transfer are called **adiabatic**; an example is a thermos or a Dewar or any other kind of "perfect" insulation. Walls allowing heat transfer are called **diathermal**. A good diathermal wall might be a metallic contact between two systems. Lastly, we call a system **closed** or **isolated** if its walls are restrictive to changes in all extensive quantities $(U, V, N_1, N_2, \dots)$.

We summarize,

Wall property	Use
rigid	to measure volume
moveable	to do mechanical work
impermeable	to measure chemical composition (mole numbers)
permeable	to change chemical composition
adiabatic	to suppress heat transfer
diathermal	to allow heat transfer

2.3.2 *Energy measurement*

For a purely mechanical system (of masses and pulleys etc.) the energy of a system, consisting of kinetic energy and potential energy, can only be changed by applying forces (gravity, electric and magnetic forces on charges, even friction) that do work on the system. In a thermodynamic system the energy can be changed by two fundamentally different methods: (a) by applying forces doing work, e.g. pushing a piston into a cylinder filled with a gas: or (b) by transferring heat into the system, e.g. by submersing the gas cylinder into a hot or a cold bath or by heating it up with a torch. To quantify this statement we assign an energy U_{in} to a system in some initial state and perform work and transfer heat. Thus the energy in the final state reads

$$U_{\text{fin}} = U_{\text{in}} + W_{i \to f} + Q_{i \to f} \qquad (2.1)$$

And the energy change to go from the initial to the final state is

$$\Delta U_{i \to f} = U_{\text{fin}} - U_{\text{in}} = W_{i \to f} + Q_{i \to f}. \qquad (2.2)$$

Here $W_{i \to f}$ is the work done on the system and $Q_{i \to f}$ the heat transfer into the system to take the system from the initial to the final state.

Remark 2.2. In non-relativistic mechanics (classical or quantum) the energy can only be specified up to a constant; the reason is simple: although the kinetic energy has an absolute zero, namely for the object at rest, the potential energy is only determined up to a constant because (i) in classical mechanics the potential energy only enters the equations of motion through its derivative with respect to space, and (ii) in quantum mechanics a constant added to the potential energy only shifts the energy scale of the eigenvalues and those are only physically relevant as energy differences in transitions.

Heat transfer is at the center of thermodynamics and sets it apart from any other physical theory.

2.3.2.1   *Work*

Work performed on a system can be mechanical, electrical, magnetic or chemical. For now let us look at the mechanical work performed by pushing a piston of area A in a cylinder by an incremental distance dz applying a force f resulting in an incremental amount of work

$$đW = f dz = (f/A)d(Az)$$
$$= -PdV \qquad (2.3)$$

where $P = f/A$ is the pressure. If the external force acting on the piston is directed inward (as it would be if we were to put an additional weight on it) then the piston will move down reducing the volume, i.e. $dV < 0$, and the force will do work on the system increasing its energy, i.e. the work is positive from the system's point of view. This sign convention in the natural sciences is opposite to that used by engineers for whom engines do positive work on another system.

Having an expression of the work for an infinitesimal change in volume we get the work to go from some initial to some final volume by integration

$$W_{i\to f} = -\int_{V_i}^{V_f} P(V)dV \qquad (2.4)$$

Recall from everyday experience that the work done to go from point A to point B depends on which path one takes, i.e. whether one climbs over a mountain top or walks around its base. The same is of course true for pushing in a piston; that is why we wrote $P = P(V)$ under the integral. Because work depends on the path there is no function $W(T, P)$ and the incremental work $đW$ is **not** a perfect differential. To keep track of this fact we put a slash through the differential.

Let us do three experiments:

(i) Let us push in the piston at constant pressure, $P = P_0$, but through a sequence of equilibrium states. For an ideal gas the equilibrium states are described by the ideal gas law which we write as

$$P = NR\frac{T}{V} \qquad (2.5)$$

As we decrease the volume, with the proviso that the pressure remains constant, we see that at the same time we have to also lower the temperature by cooling the gas, but in such a manner that the ratio T/V stays constant. This achieved we can evaluate the integral easily and get

$$W_{i \to f}(P = const) = - \int_{V_i}^{V_f} P(V)dV = -P_0(V_f - V_i) \qquad (2.6)$$

(ii) Next we push the piston in keeping the temperature constant, $T = T_0$, again through a sequence of equilibrium states. In this situation

$$P = NRT_0 \frac{1}{V} \qquad (2.7)$$

and the work done is

$$W_{i \to f}(T = const) = - \int_{V_i}^{V_f} P(V)dV = -NRT_0 \ln(V_f/V_i) \qquad (2.8)$$

(iii) Lastly, let us change the volume with the blow of a hammer. In this case the in-between states are completely out of equilibrium and we know absolutely nothing about how the pressure changes with volume, or for that matter what the temperature does. We would have to measure $P(V)$, and this would require some ingenuity, particularly if the blow is hard and fast. Such a process is outside the realm of thermodynamics and must be handled by the methods of non-equilibrium thermodynamics.

2.3.2.2 *Quasi-static processes*

Quasi-static processes are ones that are so slow that during its time evolution, the system goes through (approximately) a sequence of equilibrium states. These systems can be described and calculated by the methods of thermodynamics.

What is slow? Or, as one should always ask in science: slow compared to what? As we already discussed, every system has an internal relaxation time, t_{rel}, over which it settles into an equilibrium state. In a dilute gas, it is the time required for a density disturbance to travel a few times across the container which is roughly $t_{rel} \approx L/v_s$ where L is a linear dimension of the container and v_s is the speed of sound. As an example, for the piston in the combustion chamber of a gasoline engine with $L \approx 20$ cm and $v_s \approx 500$ m/s we get $t_{rel} \approx 0.4$ ms. Thus, if an engine is revving with less than $150,000$ rpm (which all car engines do!) the engine performance can

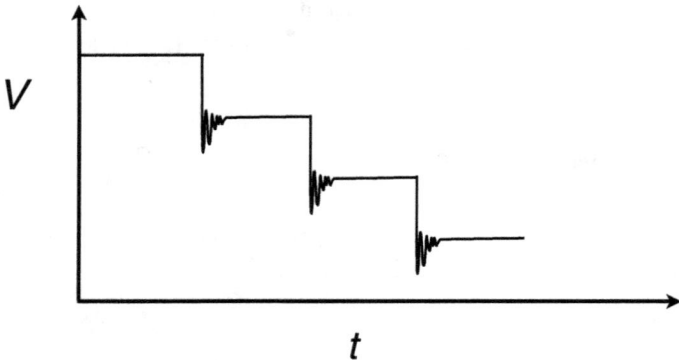

Fig. 2.2　In a quasi-static process externally induced changes in the system are slow on the time scale of intermittent fluctuations induced by these changes.

still be described by quasi-static processes because its macroscopic behavior is slow compared to its internal time scale! If you read that such processes are infinitesimally slow, it means they are slow on the time scale of internal relaxation processes.

2.3.2.3　*Heat transfer and Joule's theorem*

In a quasi-static process we obtain the total energy change from some initial equilibrium state with internal energy U_i to some final state with U_f by integrating infinitesimal steps

$$\Delta U_{i \to f} = U_f - U_i = \int_i^f dU \tag{2.9}$$

where

$$dU = đW + đQ \tag{2.10}$$

We already know how to calculate and thus measure the work done along a particular path. What about heat transfer? The surprising answer is: there is no way to either calculate or measure heat transfer directly. However, Joule showed that there always exist adiabatic walls such that the energy change $\Delta U_{i \to f}$ can be achieved by mechanical work alone. This is very important, because it reduces energy measurements, even in the presence of heat transfer, to measurements of mechanical work, something that is familiar from mechanics. Take as an example a cylinder filled with a gas and fitted with a piston. Initially the volume occupied by the gas is V_i, its pressure is P_i, and the (arbitrary) internal energy is U_i. We now heat the gas with a torch, resulting in the piston's movement to a new position such that the volume is now V_f, the pressure is P_f, and the temperature is T_f. The question is, what is U_f? Obviously $\Delta U_{i \to f}$ in this process is achieved by heat transfer which we cannot measure directly. However, we can mechanically move the piston and mechanically stir the gas with a paddle wheel to expand the gas and change its pressure and temperature. For both mechanical processes we can determine the work done, and thus we know U_f. However, Joule noted that such adiabatic walls that reduce the determination of an energy change to a measurement of mechanical work might only exist for one direction, i.e. to go from the initial state to the final state or vice versa. But that is fine, as we are ultimately interested in measuring the energy change between two equilibrium states. Example: if pushing a piston into a cylinder increases the pressure in the gas and decreases its volume, pulling it out again will reduce the pressure in the gas and increase the volume again to their initial values. On the other hand, if you used a paddle wheel to stir a liquid to increase (mechanically) its temperature, putting the paddle wheel in reverse will not cool it down again! This is a first manifestation of irreversibility; more on that later.

We summarize **Joule's theorem**:

> **For any thermodynamic system there always exist adiabatic walls that reduce the determination of the internal energy U to a measurement of mechanical work alone. However, such adiabatic processes (work) might only exist for one direction.**

An example for Joule's theorem: A gas is contained in a cylinder with a moveable piston. It is experimentally observed that if the walls are adiabatic, a quasi-static decrease in volume (pushing in the piston) results in an increase in pressure according to

$$PV^{5/3} = const \tag{2.11}$$

Such a curve is called an adiabat (see Figure 2.3 and also Problem 2.1). The constant depends on the initial point (P_A, V_A) from which we started, and can be measured. This allows us to determine the internal energy anywhere along this adiabat, because with no heat transfer possible through an adiabatic wall, we see from (2.10) that the change in the internal energy is solely due to the work done on the system alone, $dU = dW$. For example,

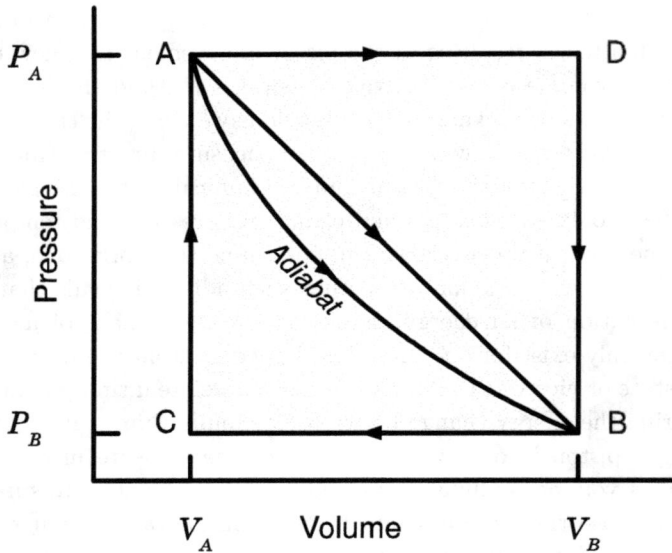

Fig. 2.3   P-V diagram indicating several processes including an adiabat.

the energy U_B at point B in Figure 2.3 is, relative to the energy at point A

$$U_B - U_A = \int_A^B dU = \int_A^B dW = -\int_{V_A}^{V_B} P(V)dV \qquad (2.12)$$

Here we have to integrate along the adiabat (2.11) because that is path our experiment was designed to follow. Along the adiabat we have $P = $ const $V^{-5/3}$. Therefore, we get

$$U_B - U_A = -\int_{V_A}^{V_B} P(V)dV = -const\int_{V_A}^{V_B} V^{-5/3}dV \qquad (2.13)$$

$$= \frac{3}{2}const(V_B^{-2/3} - V_A^{-2/3})$$

Now, along the adiabat we have

$$P_A V_A^{5/3} = P_B V_B^{5/3} = const \qquad (2.14)$$

so that we get

$$U_B - U_A = \frac{3}{2}(P_B V_B - P_A V_A) \qquad (2.15)$$

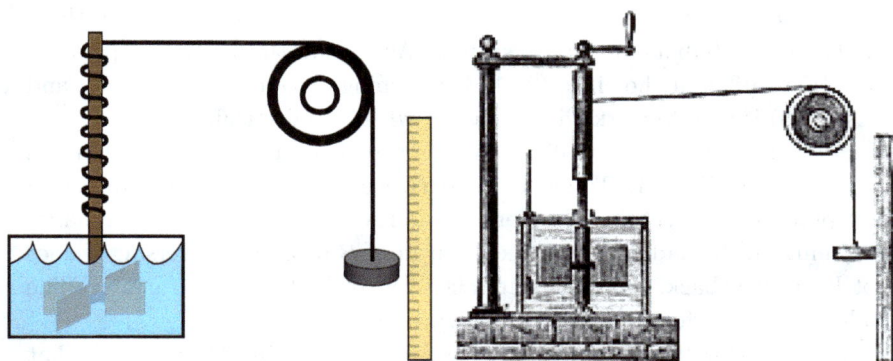

Fig. 2.4 Joule's apparatus for measuring the mechanical work equivalent of heat. A descending weight attached to a string causes a paddle immersed in water to rotate. The change in gravitational potential energy that occurs as the weight is lowered is directly related to the work that must be done to cause the wheel to rotate. Dissipative forces subsequently convert this motion into heat.

Note that in this form the change in internal energy does not depend on the constant anymore. Also note, that this change is identical to the change in an ideal gas, i.e. the system **is** an ideal gas.

To get the internal energy away from this adiabat we need a second adiabatic process. To this end we install a small paddle wheel in the cylinder, driven by an external motor exerting a torque τ and driving the paddle at an angular speed ω. As a result of the work done by the paddle wheel, the pressure in the cylinder is found experimentally to increase at a rate

$$\frac{dP}{dt} = \frac{2}{3}\frac{\omega}{V}\tau \tag{2.16}$$

The work done by turning the wheel through a small angle $d\theta$ is then

$$dW = \tau d\theta \tag{2.17}$$

or, using the previous equation

$$dW = \frac{3}{2}VdP \tag{2.18}$$

This adiabatic process connects any point along the $PV^{5/3} = const$ adiabat, obtained by doing work moving a piston, with any point above it e.g.

$$U_D = U_B + \int_B^D dW = U_E + \frac{3}{2}V_B(P_D - P_B) \tag{2.19}$$

It is important to note that work can be done by many means and there can be many adiabats for a given system. As a result the analytic expression for dW is different, i.e. it is (2.11) for pushing a piston adiabatically and it is (2.18) if you do work with a paddle wheel adiabatically.

A comment on irreversibility: pushing the piston in adiabatically brings us from point B to A. To get back to B, we simply reverse the direction and pull the piston out. This process is reversible. On the other hand, switching on the paddle wheel takes you from B to D, but reversing it does not bring you back, but rather it takes you to higher temperatures. The paddle wheel performs an *irreversible* process.

To get below the AB adiabat, e.g. to get to point C, we recall that Joule said that between any two points in the phase diagram there is always an adiabatic process that either takes you from A to C *or* from C to A. Of course, we can achieve the latter process adiabatically with the paddle wheel. We write

$$U_C = U_A - \int_C^A dW = U_A + \frac{3}{2}V_A(P_A - P_C) \tag{2.20}$$

Consequently, we have the internal energy $U = U(P, V)$ everywhere. It is by the way suggestive to use the state $P = V = 0$, i.e. no gas in the cylinder, as the reference or fiducial state with respect to which we measure internal energy so that

$$U(P, V) = \frac{3}{2}PV \tag{2.21}$$

which is nothing but the ideal gas law.

Example 2.3. Let us next, as an application of our new found knowledge, study a non-adiabatic process, namely the expansion of the gas at constant pressure from A to D. We start by calculating the work

$$W_{A \to D} = -\int_{V_A}^{V_D} P(V)dV = -P_A(V_D - V_A) \tag{2.22}$$

Note that this work is negative: the gas does work against the piston. Thus to go from A to D at constant pressure we must transfer heat, part of which is used to do this work on the surroundings. To get the heat transfer we recall (2.1) and get

$$Q_{A \to D} = U_D - U_A - W_{A \to D}$$
$$= \frac{5}{2}P_A(V_D - V_A) \tag{2.23}$$

Thus to expand the cylinder at constant pressure from A to D, we must add heat to the system and take work out of it at $2/5$ the rate. Where does the extra energy go to? Obviously to increase the internal energy of the system! Using kinetic theory we get a plausible picture: increasing the volume at constant temperature decreases the pressure because the gas particles have to travel further to collide with the walls, thus reducing the pressure. To keep the pressure constant we must heat up the gas, in other words increase the kinetic energy of the gas particles. This is done by transferring heat into the system.

2.4   Problems

Problem 2.1. *The internal energy of a dilute gas of molecules is given by* $U = cPV$ *where* $c \geq 3/2$. *Find the adiabat* $P = P(V)$. *(For a dilute gas of ultra-relativistic classical particles with speeds close to the speed of light (such as in a Supernovae)* $c = 3$.)

Hint: Along the adiabat $đQ = 0$ *so that* $dU = -P(V)dV$. *With* $P = P(V)$ *the internal energy becomes a function of* V *alone along the adiabat and* dU *can be calculated.*

Problem 2.2. *A dilute gas of molecules (see Problem 2.1) also obeys the ideal gas law* $PV = NRT$. *Such a gas is expanded isothermally from an initial state* (P_i, V_i) *to a final volume* V_f.

(a) *What is the final pressure?*
(b) *What is the amount of work* $W_{i \to f}$ *done in this expansion? Is work done on the system or by the system?*
(c) *What is the heat transfer* $Q_{i \to f}$? *Is it put into the system or extracted from it?*

Problem 2.3. *A dilute gas of molecules is expanded from an initial state* (V_A, P_A) *to* (V_B, P_B) *in such a way that the pressure rises linear with volume. It is then compressed at constant volume to* $(V_C = V_A, P_C = P_B)$. *Finally it is brought back to the initial state at constant volume.*

(a) *Calculate the amount of work* $W_{i \to f}$ *done in this expansion and the heat transfer* $Q_{i \to f}$ *for each of the three processes.*
(b) *What is the overall work and heat transfer?*

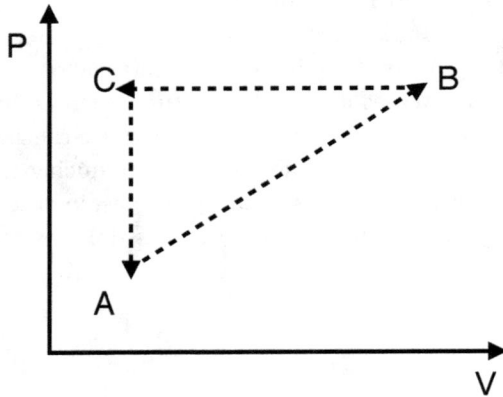

Problem 2.4. *Find the adiabat of a dilute gas of molecules in the $T-V$ plane and the $T-P$ plane.*

Problem 2.5. *The internal energy of blackbody radiation in a volume V can be written as $U = 3PV$ or $U = bVT^4$.*

(a) Find the adiabat in the $P-V$ plane.
(b) Find the adiabat in the $T-V$ plane.

Problem 2.6. *An evacuated cavity contains blackbody radiation at a temperature T.*

(a) Calculate the work done in an isothermal expansion from V_i to V_f and relate it to the change in internal energy and to the heat transfer?
(b) Is work and heat transferred into the system or extracted?

Chapter Summary

(1) Thermodynamics deals with large systems in and close to equilibrium.

(2) The equilibrium states of a large system are independent of time and of previous history.

(3) The equilibrium states are completely specified by a small set of extensive variables such as volume, mole numbers, internal energy, electric charge, electric and magnetic dipoles, etc.

(4) Walls around a system allow the measurement and changes in the extensive variables.

(5) The entropy accounts for the disorder in the system.

(6) The internal energy can be changed by doing work on the system or transferring heat.

(7) Joule's theorem allows us to reduce the measurement of heat transfer to that of work.

Chapter 3

The Laws of Thermodynamics

There are four laws of thermodynamics that have been abstracted and generalized from experimental results. Mathematically speaking they are the axioms and postulates on which the edifice of thermodynamics is erected. Fortunately, these laws are today very intuitive and plausible, although they were far from that at the beginning of thermodynamics.

3.1 Zeroth law: The fundamental relation

Frequently the status of a zeroth law is accorded to the statement *"If two systems are separately in equilibrium with a third system, they are in equilibrium with each other"*. Although this is a nontrivial statement, a more important role is played by the fact that a single function, namely the entropy (as a function of the extensive variables of a system), controls and defines all its equilibrium properties; it is therefore also called the **Fundamental Relation**. We will see shortly that the statement on the equilibrium of three systems can be derived from the first and second law of thermodynamics. For completeness, we recall that S is extensive. This implies that $S(\lambda U, \lambda V, \lambda N) = \lambda S(U, V, N)$, i.e. S is a homogeneous function of first order.

Entropy is a measure of disorder in the system. To understand this let us look at an example from everyday life.

Example 3.1. Assume we want to place n indistinguishable objects into N compartments, at most one in each. If $n = N$ we just need to make one trial and we know for certain that each compartment houses one object; the disorder is zero. If $n \neq N$ we have a total of $\Omega = \binom{N}{n}$ choices all leading to different sets of compartments being occupied by an object; disorder is large. Because entropy is introduced as a measure of disorder it must be a function of Ω, but which function $S = S(\Omega)$ is it? We do another experiment, this time with two sets of compartments A and B into which we can distribute the objects. We place n_A objects into the first set with a total of $\Omega_A = \binom{N}{n_A}$ choices and the remaining $n_B = n - n_A$ objects into the second set with $\Omega_B = \binom{N}{n_B}$ choices. The total number of choices is therefore $\Omega_{AB} = \Omega_A \Omega_B$. On the other hand, we insisted in thermodynamics that S is extensive; therefore $S(\Omega_{AB}) = S(\Omega_A \Omega_B) = S(\Omega_A) + S(\Omega_B)$. But this functional relation has only one solution, namely $S = a \ln(\Omega)$. Later on we will identify $a = k_B$ for dimensional reasons.

Remark 3.1 (history). *This is essentially the argument with which Ludwig Boltzmann (1844-1906) arrived at the connection of entropy with probability and thus explained the nature of entropy for the first time. His formula $S = k \ln W$ (W stands for Wahrscheinlichkeit = probability) is engraved on his tombstone in the Central Cemetery in Vienna (where other famous people like Ludwig van Beethoven are also honored, together over 3 million others).*

3.2   First law: Energy conservation

The energy of an isolated system is conserved. This is a cornerstone in Newtonian mechanics. For a purely mechanical system (of masses and pulleys etc.) the energy of a system consisting of kinetic energy and potential energy can only be changed by applying forces (gravity, electric, and magnetic forces on charges, even friction) that do work on the system. In a thermodynamic system the energy can be changed by two fundamentally different methods: (a) by applying forces doing work, e.g. pushing a piston into a cylinder filled with a gas: or (b) by transferring heat into the system, e.g. by submersing the gas cylinder into a hot or a cold bath, or by heating it up with a torch. To quantify this statement we recall (2.10) that a change in energy can be achieved by work or heat transfer

$$dU = dW + dQ \tag{3.1}$$

or, for a finite change,

$$U_{fin} = U_{in} + W_{i \to f} + Q_{i \to f} \tag{3.2}$$

Heat transfer is at the center of thermodynamics and sets it apart from any other physical theory. The work performed on a system can be mechanical, electrical, magnetic, or chemical.

Fig. 3.1 The headstone of a great thermodynamicist, Ludwig Boltzmann, in the Central cemetery in Vienna.

Remark 3.2. Energy conservation plays a fundamental role in all of physics. Indeed, an apparent violation of energy conservation is taken as an indication that some new phenomena have not been taken into account. Three examples should explain the point: (a) In a system of charged particles subject to electric and magnetic fields or electromagnetic radiation, the accelerating particles emit electromagnetic radiation. Thus mechanical energy is not conserved, instead one must add the energy of the electromagnetic field to obtain energy conservation. (b) In relativistic mechanics energy can be converted into mass. Thus only the sum of energy and mass satisfy energy conservation. (c) Looking at tracks of cosmic rays in a cloud chamber, the kinetic energies of decaying particles (obtained from the circular tracks in a magnetic field) suggested the presence of uncharged and very light particles called neutrinos. Again a violation of energy conservation lead to new discoveries.

Newton's law of energy conservation is obviously a bold abstraction that seems to be in contradiction to common experience. Take as an example a ball rolling on a flat plane: its initial kinetic energy will decrease with time due to friction and bring the ball to a stop. You can build an air table that reduces friction to a minimum but you can not eliminate it altogether. Thus for an ideal situation without friction Newton's energy conservation is okay. In not so ideal circumstances friction will generate heat in the ball and the table and a generalization of the concept of energy conservation is needed. Thermodynamics does this.

Three names are associated with energy conservation in thermodynamics: Rumford, Mayer and Joule. It is very instructive to follow their lines of argument to appreciate how science progresses. All three were multi-talented people with colorful backgrounds.

Remark 3.3 (history). *Count Rumford (1753-1814) was born Benjamin Thompson in Massachusetts and attended a village school. At age thirteen he was apprenticed to a merchant. He excelled at his trade and came into contact with well educated people in Boston, adopting their sophistication and conversational skills. Also at this early stage he got interested in the nature of heat, performing some initial experiments. In 1772 he married a wealthy and well-connected woman. Through her influence he was appointed a major in a New Hampshire militia. When the American Revolution began he sided with the British, which resulted in a mob attacking his house. He subsequently fled to the British. While working with their armies, he conducted experiments on the force of gunpowder. This work was published in 1781 in the Philosophical Transactions of the Royal Society (London) to great acclaim, establishing him as a scientist. In 1785 he started to work for the King of Bavaria, reorganizing his army. While overseeing the boring task of boring cannons, he measured the heat produced by the tedious work done by the horses. From this work, he published An Experimental Enquiry Concerning the Source of Heat which is Excited by Friction in 1798. This is the first account of identifying heat as a form of energy transfer and suggesting energy conservation.*

Rumford made many inventions such as the double boiler, pressure cooker, drip coffeepot, and the highly efficient Rumford fireplace, all of them still in use today. He was also active as a social worker, convinc-

ing the King that he would greatly benefit if he were to look after the poor of the city by giving them the opportunity to work. This lead to the establishment of workhouses. In a strike of genius, he also solved the accompanying task of providing food for the workers. The Rumford Soup (recipe: 1 part pearl-barley, 1 part dried yellow peas, 4 parts potato, some salt, all boiled in old, sour beer that nobody wants to drink anymore, eaten with bread) is perfect from a modern nutritional point of view, over a century before food science and nutrition came into existence. It was used as a military ration for most of the 19th and 20th centuries. In one of his projects aimed at providing work for the poor of Munich, Rumford helped to establish the English Garden. This would be the first public garden in Europe, a wonderful place to be to this day.

Remark 3.4 (history). *Julius Robert von Mayer* *(1814-1878)* grew up in southern Germany, where he studied medicine, and in 1840, became a ship's surgeon on a Dutch three-mast sailing ship for a journey to Djakarta. During his travels, Mayer observed that storm-whipped waves were warmer than the calm sea, leading him to think about "whether the directly developed heat alone or whether the sum of the amounts of heat developed in direct and indirect ways contributes to the temperature". He concluded that water must be warmed up by vibrating! Once he was in Djakarta, he made another important observation: while letting blood from sick sailors by lancing a vein, he observed that the blood was far too red, more like arterial blood. Today we know that arterial blood is redder because it carries oxygen, but that was totally unknown at Mayer's time. His conclusion: people in the tropics burn less of the food they eat. They generate less heat. On the other hand food fuels our power output, and Mayer realized that food thus must also fuel heat supply. Conclusion: energy supply through food intake is spent either on mechanical work or on heat transfer. This implies that energy is conserved. After returning to Germany he devised experiments to measure the mechanical equivalent of heat, obtaining an initial value of as 3.58 kJ/kcal. Through subsequent work, he was able to refine his value, obtaining a result of 4.168 kJ/kcal. This is in excellent agreement with today's accepted value of 4.184 kJ/kcal. Thus energy conservation was established a second time.

Remark 3.5 (history). *James Prescott Joule* *(1818-1889)* was the son of a wealthy brewer in Manchester. He was tutored at home and eventually studied with John Dalton and John Davis. He managed his father's *brewery until it was sold in 1854. His first science paper was on electricity, which he published in 1838. Five years later, he wrote his first paper on thermodynamics in which he states "the mechanical power exerted in turning a magneto-electric machine is converted into the heat evolved by the passage of the currents of induction through the coils; and, on the other hand, that the motive power of the electro-magnetic engine is obtained at the expense of the heat due to chemical reactions of the battery by which it is worked." In 1845 he published his best value of the mechanical equivalent of heat obtained with his paddle wheel experiment (see Figure 2.4) as 4.41 kJ/kcal. His work was not received well by people like Lord Kelvin, Faraday, Stokes and others. Note that Joule includes the energy of chemical reactions in his overall conservation of energy as did Mayer before him. Although Mayer and Joule, both outsiders, had difficulties convincing the scientific establishment, it was Hermann von Helmholtz in his 1847 paper that credited both for the establishment of the first law of thermodynamics. In this paper Helmholtz also formulated the second law of thermodynamics on the increase of entropy. With Helmholtz and Rudolf Clausius, thermodynamics finally was taken over by trained physicists and mathematicians.*

3.3   Second law: Entropy always rises

Let us perform an experiment: we start with a system, isolated from the rest of the world, with thermodynamic variables (U, V, N). Let us now divide the system into two subsystems with variables $U^{(1)}$, $V^{(1)}$, $N_i^{(1)}$ and $U^{(2)}$, $V^{(2)}$, $N_i^{(2)}$ such that the complete system has $U = U^{(1)} + U^{(2)}$, $V = V^{(1)} + V^{(2)}$, and $N_i = N_i^{(1)} + N_i^{(2)}$. We now make the dividing wall rigid, impermeable, and adiabatic. By moving the wall we change the volumes to $V^{(1)} - \Delta V$ and $V^{(2)} + \Delta V$. Similar manipulations will change the internal energies and the mole numbers. But now we know more about the systems, namely that the thermodynamic variables in one are higher/lower than in the other. In other words, we have decreased the entropy of the composite system by imposing an internal **constraint** on the system. Or, turning the argument around, making the dividing wall moveable, permeable, and diathermal will lead to energy and mass exchange, making the two parts indistinguishable thereby increasing the entropy. Thus we can say that as a function of all possible internal constraints on the system the entropy attains a maximum for the unconstrained system in equilibrium. Therefore a system out of equilibrium will eventually settle in a state in which the entropy is a maximum, i.e.

during the course of its evolution the entrcpy always rises.

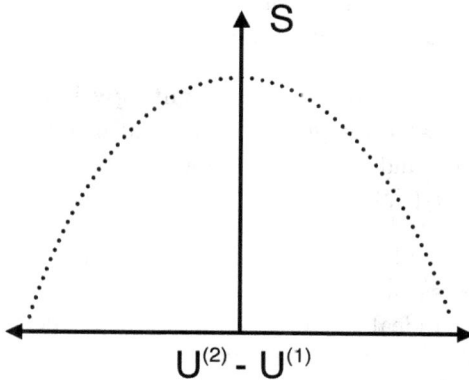

Fig. 3.2 Principle of maximum entropy over the constrained states. See Problem 3.1 for an explicit construction of this curve.

3.3.1 *Examples of the second law in physical systems*

(a) Bring a balloon filled with helium gas into a room at atmospheric pressure. Punch the balloon so that the helium gas escapes. It will mix with the air of the room until eventually it is distributed uniformly over the entire room. Whereas initially we knew that all the helium was in the balloon, in the final state it is everywhere in the room. Thus disorder has increased and the entropy has been maximized.

(b) Throw some ice cubes into a warm liquid other than water. Initially the water molecules are fixed in the ice cubes, after melting they will be anywhere in the liquid (assuming the liquid can be mixed with water). Again, disorder has increased and so has the entropy.

(c) Add some food coloring to water. Initially the color is seen in patches and streaks but after sufficient time the coloring is uniform. This is caused by diffusion of the coloring liquid and water with respect to each other. No stirring required!

(d) A solid system: add some colored sand to similar sand of a different color with same grain size and density. Nothing happens! There is a huge activation barrier for the sands to mix. You can overcome this barrier by shaking the sand, with the result that the sands mix and look homogeneous. If the two sands are of different grain sizes or densities, then the bigger or heavier grains will settle to the bottom due to gravitational pull. We then have an inhomogeneous equilibrium due to the presence of an external force field.

Corollary of the second law: the entropy is a monotonic function of the internal energy. As a consequence we can always invert the entropy function and obtain the Fundamental Relation in the energy representation $U = U(S, V, N_1, N_2, ...)$.

Remark 3.6. U is only the Fundamental Relation as a function of all extensive variables as indicated. Thus the expression we derived for the internal energy of an ideal gas, $U = \frac{3}{2} NRT$, or equivalently, $U = \frac{3}{2} PV$, are NOT fundamental relations but merely convenient expressions to calculate numerical values. Functions, by their very definition, are defined by their arguments.

3.4   Understanding the second law

There is no doubt that one of the most conceptually challenging, and at the same time revolutionary, aspects of thermodynamics is the notion that any system in isolation is driven toward a state that maximizes disorder. We will therefore elaborate this point with two detailed examples in this and the next Section. Not only does the concept of disorder appear to be somewhat subjective, this statement seems to offer no clue as to the underlying physical mechanism.

To begin, we recall a statement made by the great physicist Richard Feynman:

> If, in some cataclysm, all scientific knowledge were to be destroyed, and only one sentence passed on to the next generation of creatures, what statement would contain the most information in the fewest words? I believe it is the atomic hypothesis (or atomic fact, or whatever you wish to call it) that all things are made of atoms - little particles that move around in perpetual motion, attracting each other when they are a little distance apart, but repelling upon being squeezed into one another. In that one sentence you will see an enormous amount of information about the world, if just a little imagination and thinking are applied.

Among many other subjects, Feynman possessed a deep understanding of thermodynamics. We therefore follow closely his simple yet elegant argument for the second law of thermodynamics.

Consider a box full of particles (molecules), half of them blue (e.g. nitrogen molecules), and the other half red (say oxygen molecules). Let us assume that each particle possesses some velocity that is initially pointed in some random direction. The particles all interact with one another in an identical fashion. We shall assume they are hard spheres, but this detail is unimportant - the interaction could just as easily be that of charged particles or neutral molecules.

Let us also assume that at some point in time, when we started looking at the particles, all of the blue particles are on the left hand side of the box and all of the red are on the right. We now proceed to observe the system as it evolves through time. Since all of the velocities were originally random, initially all of the particles move in different directions. This results in collisions. For particles close to the left wall (i.e. far away from the interface that initially separates the blue from the red particles) those collisions will all involve particles of the same color. The same is true at the opposite wall. At the interface, collisions will take place between pairs of different colored particles. For some of those scattering events (another word for collision or interaction), particles will be reflected back toward their initial side of the box. For others, the particles will scatter in directions that cause them to switch sides, crossing the imaginary line that divides the box. Other particles will of course cross this line without the help of scattering, by virtue of their initial trajectories. By this mechanism, the interface between blue and red at the center of the box will become less well defined - blue particles will penetrate into red and vice versa. This process is known as mixing by diffusion.

Let us continue to observe the box, making note of the number of particles of each kind on the left and right hand side of the box respectively. It stands to reason that after some time t_0, there will be an instant when roughly the same number of blue particles are on either side of the box. The same will be true of the red. We will define "roughly" more concretely later (see Problem 3.6). Given this even distribution of particles, we can now say that the box is "mixed". A good measure of whether a system is mixed is whether or not an outside observer, knowing only the position of particles at a fixed time, can determine which side started with blue and which red.

We now consider a second box with the same initial coordinates, but a different set of randomized velocities. By the exact same line of reasoning, we expect that after some period of time, say t_1, this second box should also reach a mixed configuration (according to our definition).

Finally we consider another unmixed box, with different coordinates and velocities than the previous two. The only thing in common with the preceding cases is that all of the blue are on the left hand side, and all of the red are on the right. Again, based on the arguments laid out previously, we expect that after some time t_i (i is a label painted on the outside of each box), we expect to find the box in a mixed configuration.

Let us collect our observations so far. If we start with an unmixed box, regardless of how the atoms are arranged (coordinates or velocities), eventually this box will reach a point where it is mixed, provided we wait long enough. The question of mixing then is why does it remain in this state? Note that for each of these mixed configurations, if suddenly the velocities of all particles flipped sign, after a time t_i they would all return to their initial coordinates with their initial velocities (apart from the change in sign).

Although we have provided a physical pathway describing the two possible processes, unmixed $\rightarrow$ mixed, and mixed $\rightarrow$ unmixed, we know that in Nature it is only the latter that actually occurs. What then breaks this symmetry?

It turns out that the answer comes from the many different ways that we can arrange particles. If considering all possible arrangements of particles, we will see that the overwhelming majority can be classified as "mixed", while only a vanishing fraction are "unmixed". As a system evolves through time, it explores these different states. If there are many more states of one kind (mixed) than the other (unmixed), then the chances of being in an unmixed state approaches zero. Put another way, if you start with a mixed state, and you wait for the system to arrive at unmixed state, while in principle it should eventually get there, it will take a very long time; long on the timescale of the age of the universe.

Returning now to the second law, that an isolated system will tend to a state of maximum disorder, we see that what this means is that there is not a single disordered configuration which pulls the system toward it. Rather, it is simply that the vast majority of possible configurations are ones that are not well ordered. The system is just much more likely to be in a disordered state than any other.

3.5   Consequences of the first and second laws

As we stated earlier, conservation laws are stressed throughout the various fields of the physical sciences. As a result of this exposure, the first law of thermodynamics appears as an intuitive concept. Feynman came up with a simple mechanical model that very compellingly puts the first and second laws into perspective; his 'machine' is referred to as Feynman's ratchet. It will demonstrate that the second law of thermodynamics cannot be overcome.

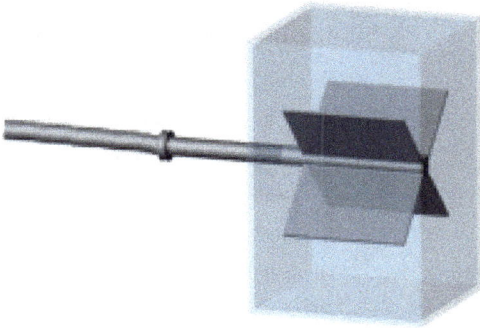

Fig. 3.3   A paddlewheel immersed in a hot fluid. Image adapted from [WikimediaCommons (2007b)].

Consider an apparatus where a paddle wheel is immersed in a hot fluid, as in Figure 3.3. When the particles (molecules of the fluid) collide with the blades, they impart a momentum. This momentum forces the axle to turn in the direction opposite to that of the incident particle. Since the movements of gas particles are random (or at least are so uncorrelated that they are essentially random), for every collision that drives the wheel in one direction there is another that causes it to reverse. Thus, in the presence of the fluid the wheel jiggles back and forth. Both its average position is unchanged and the force (exerted on the axle) is zero. No work can be extracted from the motion of the atoms using this device.

This begs the question of whether there is a different design that can convert thermal energy directly into work. We can break the symmetry that seems to prevent work from being extracted if we allow our paddle wheel to turn in only one direction. While there are different ways that this can be accomplished, for the purpose of the following arguments these details do not matter. For the sake of simplicity, we shall therefore consider the mechanical solution of the ratchet and pawl.

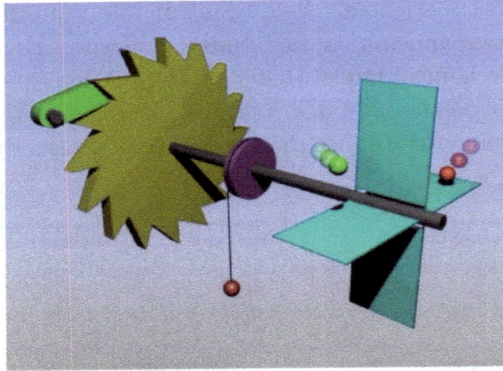

Fig. 3.4   The same paddle wheel from Figure 3.3, but with the addition of a ratchet and pawl. Image adapted from [WikimediaCommons (2008)].

In Figure 3.4, we again have a paddle wheel immersed in a hot fluid. This time the end of the axle has a toothed ratchet attached to it. When used with a pawl, the axle, and hence the paddle wheel, can turn freely in one direction but not the other. When an atom hits a paddle in the preferred direction, the ratchet advances. When one hits in the opposite direction it may not. We expect then that the force exerted on the axle, although not constant in time, should have a non-zero average value. It would seem that such a device would allow for the direct conversion of the kinetic energy of atoms into mechanical work. This is not the case. Where then is the error in our line of reasoning?

Although thermodynamics should not normally be used to consider time dependent phenomena (which are by definition out of equilibrium), in this case it is instructive to observe how this system evolves through time. Initially our device works as we had hoped - the pawl prevents the wheel from turning in one direction but allows it to turn in the other. During this initial period, work is extracted from the atomic motion. As time progresses however, an important effect (which in fact had been going on since the outset) begins to become noticeable.

To understand this effect, we must ask how exactly does the pawl force the wheel to turn only in the preferred direction? Each time the wheel turns, the pawl traces the shape of a tooth. As it passes over the peak, it snaps down onto the next tooth. What keeps it in this position? If the collision is an elastic one, we expect the pawl to bounce off the tooth, essentially having no effect on its motion. In order to "stick", energy must be dissipated somehow so that the wheel is prevented from sliding backwards. Thermodynamics tells us that this collision must mean one thing - heat is produced.

As the wheel turns and turns, the ratchet and pawl get hotter and hotter. Even if the adjoining axle is infinitely resistant to heat conduction (an impossible feat to be sure), the temperature of the pawl will increase by this process. If we rely on our everyday experience, this may not seem like something that will have any noticeable effect (apart maybe from some wear and tear). Remember though that we are discussing a device that is sensitive enough to respond to collisions by individual atoms and molecules. The fact that the ratchet and pawl are small enough to be effected by thermal fluctuations(i.e., a collision) implies that they themselves are subject to the effects of such fluctuations. Put in another way, this means that as the temperature of the pawl and ratchet increase, they begin to vibrate. These vibrations are sufficiently large to adversely affect their operation. Specifically, the pawl will start to bounce. Some of these bounces will allow the wheel to slip backwards when an atom traveling "the wrong way" collides with the paddle wheel. Other times it will actually drive the ratchet (and hence the paddles) in the opposite direction. On average however, both the average force on the axle and the position of the paddle wheel will be exactly zero; no work will be extracted. Try as we might, the second law cannot be circumvented.

Remark 3.7. In an isolated system the entropy always rises, or at best stays constant. Processes in which the entropy remains constant are called reversible. Those in which the entropy rises are called irreversible. All mechanical processes driven by conservative forces are reversible. This is a consequence of the fact that Newton's equations are invariant under time reversal, i.e. if we change time $t \rightarrow -t$ and momenta $\mathbf{p} \rightarrow -\mathbf{p}$ the system retraces its past history. This statement does not apply to mechanical systems with velocity dependent forces like friction that generate heat. The latter is an example of a system in which the entropy increases - an irreversible process cannot be reversed. Thus there is, for the evolution of macroscopic systems, an arrow of time. This apparent contradiction is resolved by Poincaré's Recurrence Theorem which states that an isolated system with conservative forces will return arbitrarily close to its initial state in a finite time. This recurrence time is huge, e.g. for a chain of harmonically coupled beads with a fundamental oscillation frequency $\omega = 10$ s^{-1} it takes about 10^{10} years to return within a few percent of its initial conditions. This is already close to the age of the universe; for larger systems the recurrence time is enormously larger. For any times that are short compared to the recurrence time the second law of thermodynamics holds true arbitrarily accurately.

Remark 3.8 (history). *Rudolf Clausius, who in 1862 introduced the concept of entropy and formulated the second law of thermodynamics, applied it to the universe as a whole and concluded that, given enough time, the universe will end in a 'heat death' of maximum entropy. Although a concept of great appeal for science fiction one must be cautious simply because the universe is not just a closed system; it might be but we do not know. In any case, the time scale for this to happen is enormous provided the universe does not collapse back onto itself as some cosmologists believe.*

Remark 3.9. Robert Emden, whom we will meet again in Section 7.3.5, circumscribed the roles of energy and entropy in a rather poetic way: "*In the huge manufactory of natural processes, the principle of entropy occupies the position of manager, for it dictates the manner and method of the whole business, whilst the principle of energy merely does the book-keeping, balancing credits and debits.*"

3.6 Third law or Nernst's theorem: Zero temperature cannot be attained

We begin historically with Nernst's theorem, which he stated as

$$\lim_{T \to 0} \Delta S = 0 \qquad (3.3)$$

This implies that as we lower temperature, we must isolate the system completely from the surrounding world to avoid even the smallest transfer of heat (which would raise the temperature again). Thus, in the quest for absolute zero, experiments become increasingly more complicated and expensive. To restate this in the form of a law,

Zero temperature cannot be attained.

This is the second formulation of the Third law, also attributable to Nernst, and is equivalent in its content to (3.3).

Note that absolute zero means there are no thermal effects in our system, and thermodynamics "reduces" in its content to classical mechanics, or, ultimately, quantum mechanics. Both of these theories are deterministic and give you complete knowledge about the behavior of a system. Thus zero temperature implies zero disorder, which means zero entropy.

In classical mechanics, the idea of a system with only conservative forces, i.e. no frictional forces, cannot be achieved completely. Likewise, a totally isolated quantum mechanical system must be in its ground state (which is unique up to symmetry-induced degeneracies). This is again an idealized world as it would require the elimination of all external electric and magnetic fields, a task which, particularly with magnetic fields, becomes exceedingly difficult the smaller the residual fields become. Once relativistic effects are included, such an isolation from the ground state electromagnetic fluctuations is in principle impossible. Likewise, gravity precludes complete isolation from the rest of the world.

The issue of degeneracy in quantum mechanics needs more attention. Degeneracies are the result of symmetries in the system. For example, consider the case of two bar magnets fixed such that they can rotate freely but are otherwise immobile. When isolated from all magnetic fields (which must be the case at $T = 0$) the magnets will align with their South and North poles in configurations **S-N-S-N** or **N-S-N-S**.[1] Both configurations are energetically equivalent. The same can be said about a macroscopic system with many such magnets all confined end-to-end in a line. The degeneracy of the system is two, and the entropy is given by $S = k_B \ln 2$. This is a nonzero value at zero temperature and is in contraction with Nernst's formulation of the third law. Not only that, this entropy is not extensive, which is in apparent violation of the zeroth law as well.

There are two ways to deal with degeneracies at absolute zero. One is Planck's re-formulation

$$\lim_{T \to 0} \frac{S}{N} = 0 \qquad (3.4)$$

where N is the number of particles (or moles) in the system. This obviously includes Nernst's formulation for a "good" extensive entropy. Another option was put forward by G.N. Lewis, namely that at absolute zero, entropy retains a constant value as other variable such as volume or mole number are changed. We conclude with a statement by Kox

---

[1]Note that in any case, the number of degeneracies in this system does not grow with system size (i.e. number of bar magnets), so in the thermodynamic limit $S/N \to 0$.

"There does seem to be a general consensus that the Third Law is of a different, less fundamental and more experimental character than the other laws of thermodynamics." – A.J. Kox

Remark 3.10 (history). *Walter Nernst (1864-1941) studied physics and mathematics after considerable soul searching as to whether he should become a professional actor. Early in his career he turned his attention to physical chemistry and electrochemistry. In 1898, at about the same time as Edison invented the lightbulb, he invented the Nernst lamp, an electric lamp using an incandescent ceramic rod. Nernst was a shrewd businessman and was able to convince AEG (a German manufacturer of all things electrical) and the city of Berlin to use his invention for street lighting - obtaining one million goldmarks, equivalent to tens of millions of dollars. And this for an invention that was never used as street lighting, since Edison's light bulb was superior. He also invented the Nernst "sock" for a gas light, still in use today for camping lights. After very careful experiments and abstraction, he established the third law of thermodynamics in 1906, which he referred to in his lectures as "my heat theorem". He received the Nobel Prize in Chemistry (1920) for this work. The story goes that he received the phone call from Stockholm sometime in the evening. Always the actor, Nernst arranged for an assistant to storm in halfway through his lecture the next morning and announce in front of the class that Nernst had just received the prize. Nernst responded with great surprise! He went on to invent the Nernst refrigerator, which was able to attain extremely low temperatures to liquify hydrogen (a machine that was so complicated that only the machinist who build it could operate it, yet Nernst managed to sell over forty), instrumentation for spectroscopy, and the electrical piano!*

3.7 Problems

Problem 3.1. *We are given a monatomic gas in equilibrium with extensive variables U, N, and V. We now insert a wall that separates the system into two equal halves, i.e. each with $U/2$, $N/2$, and $V/2$.*

(a) *Show that adding the entropies of the two halves gives indeed*
    $S(U, V, N) = S_1(U/2, V/2, N/2) + S_1(U/2, V/2, N/2)$.

(b) *Next we image that the separating wall is diathermal and impermeable but no longer rigid. Moving it, we change the volumes of the two halves into $V/2 + \Delta V$ and $V/2 - \Delta V$, i.e. we impose a constraint that negates* the fact that the two volumes are equal in equilibrium. Show that in such a virtual manipulation the entropy is changed to

$$S(U, V, N) - A(\Delta V)^2 + ...$$

Obtain an expression for the coefficient A. The entropy decreases away from equilibrium and once the constraint is removed it will increase again on the way to the equilibrium state.

(c) With a sketch, show that this behavior is also expected if U and N are constrained in the two halves.

Problem 3.2. We know that in order to compress an ideal gas, we must do work on it. By definition, this results in an increase of the internal energy of the gas. Since for an ideal gas $U = cNRT$, this implies that for a fixed mole number, T must change as a result of our work.

Consider now the case of real gases (or any other material), where the internal energy also depends on particle-particle interactions (i.e. in addition to their own kinetic energy). In such a system, is it possible to compress the material without changing its temperature? Justify your response.

Problem 3.3. Typically we think of liquids as being more disordered than solid phases of matter. This implies that liquids have more entropy. Despite this, certain materials, most notably He, are know to remain in the liquid state even at $T = 0$ K. Is this in violation of the 3rd law? Why or why not?

Problem 3.4. Is it possible for the entropy of a system to go down with time? Justify your response.

Problem 3.5. For the paddle wheel in Feynman's ratchet to move at all it must be very small and light. This might be achieved with nanotechnology. Estimate its size and moment of inertia so that a really big volatile molecule such as UF_6 at room temperature can make it spin.

Problem 3.6. Consider a room (10 m x 5 m x 2 m) full of a monoatomic ideal gas at STP. Because of fluctuations, at any given time we expect that the density of particles on either side of the room will be slightly different from one another. Assuming the motion of the atoms in the room are random, estimate the probability for the density on the left hand side to be 1% larger than the right. Will such a fluctuation ever happen? How long would you have to wait? Give an estimate (and provide justification) for the value of a fluctuation that you should expect to see over the course *of a 1 hour lecture. When such a fluctuation does occur, what will the corresponding change in pressure be?*

Chapter Summary

The laws of thermodynamics are formulated and interpreted:

0th law: the entropy is a Fundamental Relation,
1st law: the energy is conserved,
2nd law: the entropy always rises,
3rd law: absolute zero cannot be reached.

Chapter 4

Intensive Variables

To formulate the mathematical framework for thermodynamics such that one only considers quasi-static processes, we first realize that functions are not suited as a starting point because they allow for arbitrary changes, small or large. However, calculus provides us with the right tool in the form of differentials. Take any differentiable function $g(x, y)$; its differential form is given by

$$dg = \frac{\partial g}{\partial x}|_y dx + \frac{\partial g}{\partial y}|_x dy \qquad (4.1)$$

which says that an infinitesimal change from x to $x + dx$ and from y to $y + dy$ results in an infinitesimal change in the property g. In Appendix A, we have collected the essential formulae of partial differentiation and of the calculus of differentials.

A finite change is then obtained by integration

$$\Delta g_{i \to f} = g_f - g_i = \int_i^f dg \qquad (4.2)$$

Let us apply this to the Fundamental Relation in the Energy Representation, $U = U(S, V, N_1, N_2, ..., N_r)$, for a system with r components

$$dU = \frac{\partial U}{\partial S}|_{V,N_1,...,N_r} dS + \frac{\partial U}{\partial V}|_{S,N_1,...,N_r} dV + \sum_{i=1}^{r} \frac{\partial U}{\partial N_i}|_{S,V,N_1,...,N_r} dN_i \quad (4.3)$$

What is the meaning of the partial derivatives? Let us first observe that these partial derivatives are not extensive. Recall that U, V, N, S are extensive in that these variables double if you double the size of the system without changing any of its properties.

Example 4.1. Take a pot of coffee and pour it into four cups. The volume, energy, and mole number in each cup is $V/4$, $U/4$, $N/4$, adding up of course to what was in the pot because U, V, and N are extensive. The temperature T of the coffee does not change from pot to cups and back to the pot. T is intensive.

To see this formally let us look at a system that differs in size by a factor λ, i.e. we have, due to extensivity, $U_\lambda = \lambda U$, $V_\lambda = \lambda V$, $N_{\lambda i} = \lambda N_i$. For the derivatives, this implies

$$\frac{\partial \lambda U}{\partial \lambda V}|_{\lambda S, \lambda N_1, \ldots, \lambda N_r} = \frac{\lambda \partial U}{\lambda \partial V}|_{\lambda S, \lambda N_1, \ldots, \lambda N_r}$$

$$= \frac{\partial U}{\partial V}|_{S, N_1, \ldots, N_r} \qquad (4.4)$$

because whether we keep λS or S constant does not matter. Thus this derivative does not depend on the size of the system. Such variables are called **intensive**. Temperature and pressure are two examples.

4.1   Pressure

Let us first look at a change in the internal energy due to a volume change only, i.e. $dV \neq 0$ and $dS = dN_i = 0$. In this case we are doing mechanical work only to change its volume. This of course demands a force or a pressure. We get

$$dU = \frac{\partial U}{\partial V}|_{S, N_1, \ldots, N_r} dV$$

$$= \đ W$$

$$= -PdV \qquad (4.5)$$

This identifies the (negative) partial derivative of internal energy with respect to volume as the (negative) pressure

$$P = -\frac{\partial U}{\partial V}|_{S, N_1, \ldots, N_r} \qquad (4.6)$$

Remark 4.1. Again we emphasize that the pressure can only be calculated as a derivative of the internal energy if all arguments in the latter are extensive. Otherwise it is not a fundamental relation. To make this point clear, consider the following. If you were using the parametrization of the internal energy of an ideal gas $U = \frac{3}{2} NRT$, a derivative with respect to volume would give identically zero for the pressure. If instead you were using $U = \frac{3}{2}PV$, you would get $P = -\frac{3}{2}P$, which is also nonsense. In either case, you cannot keep S constant as it does not appear anywhere. To do this correctly, we first require knowledge of the fundamental relation $U = U(S, V, N)$ of a one-component ideal gas (which we will discuss shortly).

4.2 Temperature

Let us next look at a situation where there are no changes in volume or mole numbers

$$dU = \frac{\partial U}{\partial S}|_{V, N_1, \ldots, N_r} dS \tag{4.7}$$

Recall the first law of thermodynamics, which states $dU = dQ - PdV$. With no change in volume there is no mechanical work. Thus the term above must be accounting for heat transfer, dQ. It is reasonable to assume that the derivative $\partial U / \partial S$ has something to do with temperature. We will proceed by tentatively identifying it as the temperature

$$T = \frac{\partial U}{\partial S}|_{V.N} \tag{4.8}$$

although more arguments are needed to confirm this.

Because temperature is by definition a positive quantity, we conclude that U must be a monotonically increasing function of S, and similarly S as a function of U, when all other extensive variables are kept constant.

4.3 Chemical potentials

Finally, we look at a situation where only the mole number of one component changes

$$dU = \frac{\partial U}{\partial N_i}|_{S, V, N_1, \ldots, N_r} dN_i$$
$$= \mu_i dN_i \tag{4.9}$$

and call the derivative the chemical potential of the ith chemical component. It is a measure of how much a chemical reaction contributes to the free energy.

In summary, the differential form of the fundamental relation in the energy representation reads

$$dU = TdS - PdV + \sum_i \mu_i dN_i \tag{4.10}$$

Note that the partial derivatives of a function of several variables are also functions of these variables. Thus we have

$$T = T(S, V, N_1, N_2, ..., N_r) \tag{4.11}$$

$$P = P(S, V, N_1, N_2, ..., N_r) \tag{4.12}$$

$$\mu_i = \mu_i(S, V, N_1, N_2, ..., N_r) \tag{4.13}$$

These equations are called the thermal, mechanical, and chemical equations of state, respectively. For a system with r components there are $(r+2)$ equations of state. By construction, the fundamental relation in its *differential* form plus all equations of state are equivalent mathematically to the fundamental relation in *functional* form.

Temperature and pressure are familiar concepts from everyday experience, but what is the chemical potential? Looking at the differential for the internal energy at constant S and V we get

$$dU = \sum_i \mu_i dN_i \tag{4.14}$$

Thus by changing the amount of component "i", we see that μ_i is the change in internal energy resulting from adding one mole of that component with S, V, and all other $N_j (j \neq i)$ constant.

4.4 Intensive variables in the entropy representation

Before we elaborate further on the identification of temperature, pressure and chemical potentials and their significance let us briefly look at what happens when we start from the fundamental relation in the entropy representation, $S = S(U, V, N_1, N_2, ..., N_r)$, for a system with r components. Its differential reads

$$dS = \frac{\partial S}{\partial U}|_{V,N_1,...,N_r} dU + \frac{\partial S}{\partial V}|_{U,N_1,...,N_r} dV + \sum_{i=1}^{r} \frac{\partial S}{\partial N_i}|_{U,V,N_1,...,N_r} dN_i \tag{4.15}$$

Using the rules of partial differentiation we see that

$$\frac{\partial S}{\partial U}|_{V,N_1,...,N_r} = \frac{1}{\frac{\partial U}{\partial S}|_{V,N_1,...,N_r}} = \frac{1}{T} \tag{4.16}$$

$$\frac{\partial S}{\partial V}|_{U,N_1,...,N_r} = -\frac{\frac{\partial U}{\partial V}|_{S,N_1,...,N_r}}{\frac{\partial U}{\partial S}|_{V,N_1,...,N_r}} = \frac{P}{T} \tag{4.17}$$

$$\frac{\partial S}{\partial N_i}|_{U,V,N_1,...,N_r} = -\frac{\frac{\partial U}{\partial N_i}|_{S,V,N_1,...,N_r}}{\frac{\partial U}{\partial S}|_{V,N_1,...,N_r}} = -\frac{\mu_i}{T} \tag{4.18}$$

and we get

$$dS = \frac{1}{T}dU + \frac{P}{T}dV - \sum_{i=1}^{r} \frac{\mu_i}{T}dN_i \qquad (4.19)$$

together with the equations of state in the entropy representation

$$\frac{1}{T} = \frac{1}{T}(U, V, N_1, N_2, ..., N_r) \qquad (4.20)$$

$$\frac{P}{T} = \frac{P}{T}(U, V, N_1, N_2, ..., N_r) \qquad (4.21)$$

$$\frac{\mu_i}{T} = \frac{\mu_i}{T}(U, V, N_1, N_2, ..., N_r) \qquad (4.22)$$

These are completely different functions from the equations of state in the energy representation as they depend on different arguments. However, the two sets of equations of state are completely equivalent in their physical content.

Remark 4.2. A word on the choice of the measure for temperature: for historical reasons temperature was introduced to be higher for hotter objects. From this, we have the identification of T as $T = \partial U/\partial S|_{V,N}$, and that zero temperature is absolutely cold. One could have instead introduced a temperature scale $T' = \partial S/\partial U|_{V,N} = T^{-1}$. In this case, $T' = 0$ would be absolutely hot while absolutely cold would be at $T' = \infty$. Within this scheme, the notion that absolute cold is unattainable seems more natural— we can imagine that going to infinity is not easy. There is a third definition which has been contemplated over the years, namely to introduce a scale $T'' = A\ln(T)$ which would run from minus infinity to plus infinity. Again, nothing substantial would be gained except for some easier intuition. Thus we stick to T as everyone else does.

Whether one works in the energy or the entropy representation is largely a matter of convenience, and more often of simplicity.

Example 4.2. Let us figure where the equation $PV = NRT$ and $U = \frac{3}{2}NRT$ fit into this general scheme. Because neither equation contains the entropy they cannot be the equations of state in the energy representation. However, if we write

$$\frac{P}{T} = \frac{NR}{V} \qquad (4.23)$$

we see that this is the mechanical equation of state in the entropy representation; it is rather simple because it does not depend on the internal energy. Likewise, we write the second equation

$$\frac{1}{T} = \frac{3}{2}\frac{NR}{U} \tag{4.24}$$

and get the thermal equation of state in the entropy representation. Let us now simplify the problem slightly without any loss of generality by assuming that the ideal gas is enclosed by impermeable walls so that its mole number is conserved. We can then introduce reduced variables, namely the internal energy per mole $u = U/N$, the volume per mole $v = V/N$, and the entropy per mole $s = S/N$. We get

$$ds = \frac{1}{T}du + \frac{P}{T}dv \tag{4.25}$$

$$= \frac{3R}{2}\frac{du}{u} + R\frac{dv}{v} \tag{4.26}$$

Because the two terms do not depend on each other we can integrate them separately to obtain the entropy of an ideal gas

$$s = s_0 + R\ln\left[(\frac{u}{u_0})^{3/2}(\frac{v}{v_0})\right] \tag{4.27}$$

Remark 4.3. Re-introducing the mole number we have

$$S = Ns_0 + NR\ln\left[(\frac{U}{U_0})^{3/2}(\frac{V}{V_0})(\frac{N}{N_0})^{-5/2}\right] \tag{4.28}$$

Here we have introduced a reference state with values (U_0, V_0, N_0) with respect to which we do our measurement. This is necessary because the argument of a log-function (or any other) must be dimensionless.

Is this an acceptable entropy function?

(i) It is a continuous and differentiable function of the right variables (U,V,N). ✓
(ii) It is a monotonic function of U. ✓
(iii) Third law: Does it go to zero at zero temperature? **NO!**

Let us take as our reference state $T = 0$ so that $U_0 = U(T = 0)$. Thus for fixed N and V the entropy goes to a constant not equal to zero. What is wrong? As a matter of fact, thermodynamics is at its best here because it forces us to think about the physics that went into the construction of the ideal gas model: we said earlier that in general the internal energy has two contributions, kinetic energy of motion and potential energy of interactions between the gas particles. At low densities and high temperatures the contribution from the potential energy is negligible compared to the kinetic energy. However, as we cool down the gas its kinetic energy diminishes to the point where the potential energy becomes no longer negligible and eventually dominant. In a real system, interactions between gas particles are mostly attractive. This leads to liquefaction and eventually solidification: water vapor at atmospheric pressure above $100°$ C goes to liquid water below this temperature and to ice at $0°$ C. These all-important phenomena are, by construction, ignored in the ideal gas model and the laws of thermodynamics signal the fact that we have overlooked some important physics.

Having the entropy function for the ideal gas we can invert it and get its internal energy

$$U = U_0 \left(\frac{V}{V_0}\right)^{2/3} \left(\frac{N}{N_0}\right)^{5/3} \exp\left(\frac{2}{3}\frac{S}{NR}\right) \tag{4.29}$$

from which we can calculate its chemical potential

$$\mu(s,v) = \frac{\partial U}{\partial N}\Big|_{S,V}$$

$$\exp[\mu/RT] = \frac{v_0}{v}\left(\frac{u_0}{u}\right)^{3/2} \tag{4.30}$$

Remark 4.4. Recall that in an adiabatic process no heat is exchanged. If the process is reversible, the entropy is constant, implying

$$\left(\frac{u}{u_0}\right)^{3/2}\left(\frac{v}{v_0}\right) = const. \tag{4.31}$$

or, replacing $u = 3/2PV$

$$PV^{5/3} = const \tag{4.32}$$

along the path of an adiabatic process in the $P - V$ plane. And replacing $V = NRT/P$ we get in the $P - T$ plane

$$PT^{-5/2} = const \tag{4.33}$$

This is consistent with the experimental result obtained by Joule, see (2.11).

4.5   More on the physical significance of intensive variables

To establish further facts about the physical significance of the intensive variables let us consider two systems that are marginally different from each other in their energy but have equal volumes and mole numbers. Let us further assume that the walls of each subsystem are rigid and impermeable. The total energy is conserved $U = U^{(1)} + U^{(2)}$ so that

$$dU = dU^{(1)} + dU^{(2)} = 0 \qquad (4.34)$$

As we bring the two subsystem into thermal contact they will evolve towards a new equilibrium with higher and maximal entropy so that the change in total entropy

$$dS = dS^{(1)} + dS^{(2)} > 0 \qquad (4.35)$$

$$= \frac{1}{T^{(1)}} dU^{(1)} + \frac{1}{T^{(2)}} dU^{(2)} > 0 \qquad (4.36)$$

$$= \left[\frac{1}{T^{(1)}} - \frac{1}{T^{(2)}} \right] dU^{(1)} > 0 \qquad (4.37)$$

must always be positive. Let us assume that initially $T^{(2)} > T^{(1)}$ then $dU^{(1)} > 0$ must be positive, i.e. energy must flow from the hotter to the colder subsystem. As equilibration proceeds, enough energy will have been exchanged and equilibrium is established, i.e. the entropy is maximal, $dS = 0$, and the temperature is uniform throughout the combined system establishing thermal equilibrium. We can say that a gradient in temperature leads to heat transfer or heat conduction.

Remark 4.5. Because temperature is intrinsically a positive quantity a physically acceptable entropy function must be a monotonically increasing function with respect to energy (at constant volume and mole numbers) as we argued before.

Next we look at two systems separated by a fixed wall with marginally different volumes, but are otherwise identical, i.e. same composition and same internal energy. The total volume $V = V^{(1)} + V^{(2)}$ is then fixed so that

$$dV = dV^{(1)} + dV^{(2)} = 0 \qquad (4.38)$$

As we make the separating wall moveable, the systems will evolve towards a new equilibrium with higher and maximal entropy so that the change in total entropy

$$dS = dS^{(1)} + dS^{(2)} > 0$$
$$= \frac{P^{(1)}}{T} dV^{(1)} + \frac{P^{(2)}}{T} dV^{(2)} > 0$$
$$= \left[\frac{P^{(1)}}{T} - \frac{P^{(2)}}{T} \right] dV^{(1)} > 0 \qquad (4.39)$$

must always be positive. Let us assume that initially $P^{(2)} > P^{(1)}$ then $dV^{(1)} < 0$ must be negative, i.e. the wall will move to reduce the pressure in "2" by increasing its volume. As equilibration proceeds, the entropy becomes maximal, $dS = 0$, and the pressure is uniform throughout the combined system establishing mechanical equilibrium.

Finally we look at two systems separated by a wall, identical in volume, internal energy and chemical composition except for one component. The total mole number of this component $N_i = N_i^{(1)} + N_i^{(2)}$ is then fixed so that

$$dN_i = dN_i^{(1)} + dN_i^{(2)} = 0 \qquad (4.40)$$

As we make the separating wall permeable to component "i" a new equilibrium with higher and maximal entropy so that the change in total entropy

$$dS = dS^{(1)} + dS^{(2)} > 0$$
$$= -\frac{\mu_i^{(1)}}{T} dN_i^{(1)} - \frac{\mu_i^{(2)}}{T} dN_i^{(2)} > 0$$
$$= -\left[\frac{\mu_i^{(1)}}{T} - \frac{\mu_i^{(2)}}{T} \right] dN_i^{(1)} > 0 \qquad (4.41)$$

must always be positive. Let us assume that initially $\mu_i^{(2)} > \mu_i^{(1)}$ then $dN_i^{(1)} > 0$ must be negative, i.e. some of the material of the ith component will permeate through the wall from "2" to "1" to reduce the mole number in "1" and its chemical potential. As equilibration proceeds, the entropy is maximal, $dS = 0$, and the chemical potential is uniform throughout the combined system together with all other intensive variables establishing chemical equilibrium. Thus we can say that a gradient in the chemical potential leads to mass transport or diffusion.

We note that consideration of the physical significance of intensive variables leads quite naturally to the concepts of non-equilibrium transport phenomena such as heat conduction and diffusion.

4.6 Euler equation and Gibbs-Duhem relation

The mathematical statement that S, U, V, and N_i are extensive variables

$$U(\lambda S, \lambda V, \lambda N_i) = \lambda U(S, V, N_i) \qquad (4.42)$$

signals the fact that the fundamental relation is a homogeneous function of first order. Because λ is an arbitrary parameter we can differentiate both sides of this equation with respect to λ

$$\frac{\partial U(\lambda S, \lambda V, \lambda N_i)}{\partial(\lambda S)} \frac{\partial(\lambda S)}{\partial \lambda} + \frac{\partial U(\lambda S, \lambda V, \lambda N_i)}{\partial(\lambda V)} \frac{\partial(\lambda V)}{\partial \lambda}$$
$$+ \sum_i \frac{\partial U(\lambda S, \lambda V, \lambda N_i)}{\partial(\lambda N_i)} \frac{\partial(\lambda N_i)}{\partial \lambda}$$
$$= U(S, V, N_i) \qquad (4.43)$$

or

$$TS - PV + \sum_i \mu_i N_i = U \qquad (4.44)$$

This is called the Euler equation, after the famous mathematician Leonard Euler, who first wrote about homogeneous functions. Note that T, P, and μ_i are of course functions of S, V, and N_i.

To make use of the Euler equation in thermodynamics, we take its complete differential

$$dU = TdS + SdT - PdV - VdP + \sum_i \mu_i dN_i + \sum_i N_i d\mu_i \qquad (4.45)$$

and recall the differential form of the fundamental relation in the energy representation

$$dU = TdS - PdV + \sum_i \mu_i dN_i \qquad (4.46)$$

Subtracting the two we get

$$SdT - VdP + \sum_i N_i d\mu_i = 0 \qquad (4.47)$$

This is called the Gibbs-Duhem relation. To see its significance let us look at a one-component system and re-write it as

$$d\mu = -sdT + vdP \qquad (4.48)$$

in terms of molar entropy and volume. It says that the chemical potential is NOT an independent intensive variable but is a function of temperature and pressure

$$\mu = \mu(T, P) \tag{4.49}$$

This relation, obtained by applying some simple Euler mathematics about homogeneous functions, has remarkable implications for thermodynamics. Recall that to describe a simple system we introduced three extensive variables, for instance in the energy representation the entropy, the volume, and the mole number. By differentiation we got three intensive quantities: temperature, pressure, and chemical potential. What the Gibbs-Duhem relation says is that only two of them are independent. In other words, if we want to manipulate the thermodynamic state of a system we can achieve all and arbitrary changes by just varying two intensive variables, namely temperature and pressure. In other words, there are only two thermodynamic degrees of freedom for a one-component system. Similarly, if there are r chemical components in the system there are only $(r + 1)$ independent degrees of freedom to manipulate.

Example 4.3. For an ideal gas we have

$$\mu/RT - \mu_0/RT_0 = \ln\left[\frac{P}{P_0}\left(\frac{T_0}{T}\right)^{5/2}\right] \tag{4.50}$$

This gives the change in chemical potential as we move away from the reference or fiducial state, e.g. STP. We have argued that the choice of a fiducial state is arbitrary, but once chosen must be adhered to. It turns out that this arbitrariness could not be pinned down until the arrival of quantum mechanics. It was shown that associated with a particle velocity v is a wavelength $\lambda = \hbar/mv$. So for a particle moving with the average thermal velocity we get the thermal wavelength $\lambda_{th} = (h^2/2\pi mk_BT)^{1/2}$. This is an intrinsic length scale. Let us therefore take as a fiducial volume of an ideal gas at temperature T_0 a volume $\lambda_{th}^3(T_0)$ with internal energy and entropy calculated for one mole. We can then write the chemical potential as

$$\exp[\mu/RT] = N_A\lambda_{th}^3(T)\frac{N}{V} \tag{4.51}$$

4.7 Problems

Problem 4.1. *One mole of a monatomic gas is prepared at room temperature and atmospheric pressure (STP).*

(a) *What is its internal energy and what is the volume occupied?*

(b) *The temperature is raised to $100°$ C and the pressure is increased ten-fold. What is the internal energy and the volume occupied?*

(c) *What is the change in entropy going from (a) to (b)? Express your answer in terms of the universal gas constant R.*

(d) *What is the change in chemical potential going from (a) to (b)?*

(e) *What is the absolute chemical potential as given by quantum mechanics (4.51) for a gas of helium at STP? What happens to the chemical potential as the mass of the "particles" becomes infinite (e.g., for a Cadillac compared to a helium atom)?*

Problem 4.2. *The entropy of N moles of a real (not ideal) gas has been determined to be*

$$S = N s_0 + N R \ln[(v - b)(u + a/v)^c] \qquad (4.52)$$

where $v = V/N$, $u = U/N$ and s_0 is a constant.

(a) *Is this an acceptable Fundamental Relation? (Hint: check for extensivity, monotonicity in U, differentiability, and whether S vanishes at $T = 0$)*

(b) *Obtain $U = U(S, V, N)$.*

(c) *Obtain the three equations of state from U.*

(d) *Obtain the pressure as a function of temperature and volume.*

(e) *What is μ/RT at $T = 90$ K and atmospheric pressure for $a = 0.138 \times 10^{-6}$ Pa $\times$ m^6, $b = 32.6 \times 10^{-6}$ m^3 and $c = 5/2$, appropriate for molecular oxygen?*

Problem 4.3. *Two ideal gases with internal energy $U = cNRT$ and $PV = NRT$ are separated by a rigid, adiabatic, and impermeable wall isolated them from each other. Their initial values are T_1, V_1, N_1 and T_2, V_2, N_2.*

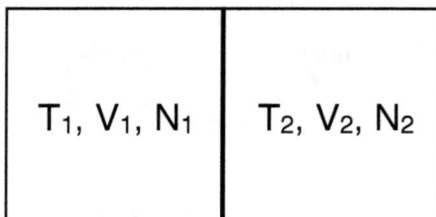

$$\boxed{\quad T_1, V_1, N_1 \quad | \quad T_2, V_2, N_2 \quad}$$

(a) *Thermal contact is established between the two systems by making the wall diathermal. After equilibrium is established what are the new values for T_1, T_2, P_1, P_2?*

(b) *Next the separating wall is made moveable but still adiabatic and impermeable. After equilibrium is established what are the new values for T_1, T_2, P_1, P_2, V_1, V_2?*

(c) *Next the separating wall is made moveable and diathermal but still impermeable. After equilibrium is established what are the new values for T, P_1, P_2, V_1, V_2?*

(d) *Next the separating wall is made permeable but still diathermal and rigid. After equilibrium is established what are the new values for T, P_1, P_2, V_1, V_2, N_1, N_2?*

(e) *Finally the separating wall is made permeable, diathermal, and moveable. After equilibrium is established what are the new values for T, P_1, P_2, V_1, V_2, N_1, N_2? What are the chemical potentials μ_1 and μ_2?*

Chapter Summary

(1) Intensive variables temperature, pressure, and chemical potentials are introduced as partial derivatives of the energy or the entropy.

(2) The Euler equation and the Gibbs-Duhem relation are derived. The latter shows that the chemical potential for a one-component system depends only on T and P.

Chapter 5

Simple Systems

In this chapter we want to look at the thermodynamics of a number of specific systems such as mixtures of gases, reacting gases, blackbody radiation, simple polymers, surface adsorbates, and magnetic systems. We begin by introducing some relevant thermodynamic quantities.

5.1 Second derivatives: expansion coefficient, compressibility, heat capacity, and more

So far we have introduced the fundamental relations in the entropy and the energy representations and argued that in quasi-static processes we should always begin by considering differential increments in all extensive variables. In this context, pressure, as an example, is a measure of the incremental increase in internal energy with a change in volume (at fixed S and N). In practical situations, most of the time we are not interested in what the incremental increase in the internal energy or the entropy are but rather what volume changes result from a pressure or temperature change, as these are the quantities most amenable to measurement. Let us therefore define

$$\alpha = \frac{1}{V} \frac{\partial V}{\partial T}\Big|_{P,N} \qquad (5.1)$$

as the isobaric (equal pressure) thermal expansion coefficient. It describes thermal expansion at constant pressure and mole number. This is obviously a very important quantity when designing an apparatus or a machine. Imagine that your task is to select the material for a piston in the combustion chamber of a car engine. It must fit snugly to get compression, but it must do so both when the engine is cold and when it is hot. Obviously you must select a material that either expands little or one that expands at the same rate as the walls of the combustion chamber. Choosing the same material would be a quick answer but not necessarily the best as the mechanical stress on the moving piston is much greater than on the walls.

Continuing with the same example, as the piston compresses the gas in the combustion chamber the pressure rises tremendously. You must make sure that neither piston nor walls change too much as a consequence of the pressure rise. Thus the isothermal (equal temperature) compressibility

$$\kappa_T = -\frac{1}{V}\frac{\partial V}{\partial P}\Big|_{T,N} \tag{5.2}$$

is a quantity of interest. We added a minus sign because (almost all) systems reduce their volume with increasing pressure. A certain situation might actually need knowledge about isentropic (iso=equal, entropic, i.e. constant entropy) compressibility for compression without adding or extracting heat from the system as you have to do for isothermal compression. In that case we need to know

$$\kappa_S = -\frac{1}{V}\frac{\partial V}{\partial P}\Big|_{S,N} \tag{5.3}$$

Also of interest is a measure for the amount of heat one must add to a system to raise its temperature. A quantity like dQ/dT would be useful. A proper definition starts from the fact that $dQ = TdS$, so that we get

$$c_V = \frac{T}{N}\frac{\partial S}{\partial T}\Big|_{V,N} \tag{5.4}$$

as the heat capacity per mole or specific heat at constant volume. Similarly one can define a specific heat at constant pressure

$$c_P = \frac{T}{N}\frac{\partial S}{\partial T}\Big|_{P,N} \tag{5.5}$$

The ratio of the specific heats is frequently denoted by

$$\gamma = \frac{c_p}{c_V} \tag{5.6}$$

There are obviously many more such partial derivatives. The examples chosen here, however, have a common feature, namely they are second derivatives of the fundamental relation. We can show this, for example, for the isobaric expansion coefficient by writing the derivative as

$$\frac{\partial V}{\partial T}\Big|_{P,N} = \frac{1}{\frac{\partial T}{\partial V}\Big|_{P,N}} \tag{5.7}$$

but $T = \partial U/\partial S|_{V,N}$ so that

$$\frac{\partial T}{\partial V}\Big|_{P,N} = \frac{\partial}{\partial V}\left[\frac{\partial U}{\partial S}\Big|_{V,N}\right]\Big|_{P,N} = \frac{\partial^2 U}{\partial V \partial S} \tag{5.8}$$

There are many such second derivatives, and each corresponds to a different experimental situation. As an example, to measure the isothermal compressibility one must ensure that temperature stays constant (by immersing the system into a larger reservoir with fixed temperature), whereas for the measurement of the isentropic compressibility you must ensure that the system is sufficiently isolated against heat transfer to and from the outside. How many such second derivative quantities do you have to measure in order to have complete knowledge of the system, i.e. to make predictions for other experiments or properties? Thermodynamics gives you the tools to reduce, for a simple system, all second derivatives to just three which are conventionally chosen to be α, κ_T, and c_P. We will learn how to do this elegantly when we study Maxwell relations in Chapter 7.

5.2 Mixture of ideal gases

As a preliminary to dealing with mixtures of different gases, let us first recall what we know about a single component gas. The fundamental relation for a dilute (ideal) gas in the entropy representation (4.28) is

$$S = Ns_0 + NR\ln\left[\left(\frac{U}{N}\frac{N_0}{U_0}\right)^c\left(\frac{V}{N}\frac{N_0}{V_0}\right)\right] \tag{5.9}$$

where so far we have set $c = 3/2$ (or equivalently $\gamma = 5/3$). This is only true for an atomic gas such as the noble gases helium, neon, argon, krypton, and xenon. For a molecular gas, $c > 3/2$ to account for the internal rotations and vibrations of the molecules. By differentiating with respect to the extensive variables we get the equations of state

$$\frac{P}{T} = \frac{NRT}{V} \tag{5.10}$$

$$\frac{1}{T} = c\frac{NR}{U} \tag{5.11}$$

Remark 5.1. For an atomic gas $c_V = 3R/2$ so that $c = 3/2$. For each of the three translational degrees of freedom of an atom (treated as a mass point), the contribution to the specific heat is $\frac{1}{2}R$. This is also the case for the center of mass motion of a molecule. However a molecule consisting of n atoms has $3n$ degrees of freedom, of which only three refer to its center of mass motion. But molecules can rotate and vibrate storing additional energy. Let us first look at a diatomic molecule such as carbon monoxide. It can rotate around an axis perpendicular to the C-O axis and through its center of mass. A rotation around an axis through the two atoms is not possible as the atoms are treated as mass points. However, the first rotation is actually degenerate as there are two rotational axis perpendicular to each other. In nonlinear multi-atom molecules there are three rotational axes. The rest of the $3n$ degrees of freedom are taken up by vibrations. CO has just one such vibration. According to quantum mechanics molecular vibrations and rotations are quantized, and as a result are excited as temperature is raised. To quantify this statement we need an estimate of a characteristic "temperature" for molecular vibrations and rotations. Molecular vibrations have typical frequencies of $\nu_{vib} \approx 10^{13} - 10^{14}$ Hz leading to a characteristic vibrational temperature of $T_{vib} \approx h\nu_{vib}/k_B \approx 500 - 5000$ K; for O_2 it is 2230 K and for H_2 it is 6100 K. In other words, at room temperature molecular vibrations are rarely excited and do not contribute much to the specific heat. With rotations it is a different matter. The characteristic temperature is given in terms of the moment of inertia and is 2 K for O_2. At room temperature it is fully excited and contributes another $\frac{1}{2}R$ for each rotational axis for a total of R. For the gases in the air at room temperature $c_V = 5/2R$. At temperatures high enough to excite vibrations c_V approaches $7/2R$ and at temperatures high enough to dissociate the molecule it drops to $2 \times 3/2R$ for the translational degrees of freedom of two atoms, see Figure 5.1.

For an ideal gas the second derivatives are

$$\alpha = \frac{1}{T}$$
$$\kappa_T = \frac{1}{P}$$
$$c_V = cR$$
$$c_P = (c+1)R \tag{5.12}$$

Interestingly, Julius von Mayer (see Remark 3.4), even before thermodynamics was developed into a complete theory, was able to derive the fact that $c_P - c_V = R$ for a dilute gas. At first sight it seems odd that thermal expansion, measured by α, is smaller at higher temperature. After all, at higher temperature the molecular motion is more energetic and thus it is easier to push the piston out. However, always note what is kept constant when taking a partial derivative. In this case it is the pressure. Thus as we increase temperature the volume must increase, while particle density, N/V, and with it the particle flux hitting the piston is reduced. A similar argument applies to the compressibility κ_T. Beware that it is only for the

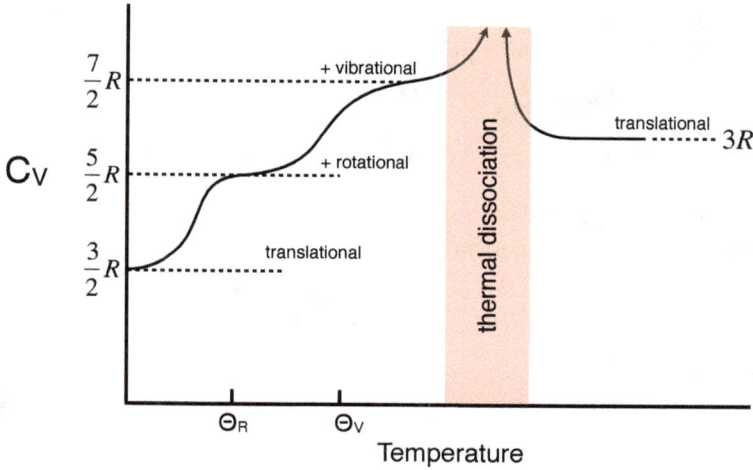

Fig. 5.1 Schematic of the specific heat for a diatomic molecule. As temperature is increased, higher energy excitations (rotations and vibrations) become accessible. This effectively increases the number of degrees of freedom of the system, and hence the specific heat. At sufficiently high temperatures, the molecule is no longer bound—dissociation occurs and the number of species doubles. Each atom possesses only translational freedom however, hence the specific heat is simply 3R.

case of an ideal gas that these quantities have such simple forms, in general they depend on T and P in complicated ways.

A natural approach to considering a mixture of gases is to begin with a homogeneous system which we divide into two parts by inserting an artificial wall. We separate the volume V occupied by N moles into two volumes V_1 and V_2. We then have $N_1/V_1 = N_2/V_2 = N/V$ and $U_1/N_1 = U_2/N_2 = U/N$, and the total entropy is indeed $S = S_1 + S_2$. Note that inserting and removing the dividing wall is a reversible process requiring neither work nor heat transfer.

Next we consider two *different* ideal gases in separate vessels of the same volume V and at the same temperature. Because the gases do not interact with each other, their total entropy and internal energy are the sums of the two

$$S(U, V, N_1, N_2) = S_1(U_1, V, N_1) + S_1(U_2, V, N_2) \tag{5.13}$$
$$U = U_1 + U_2 \tag{5.14}$$

Because we keep the temperature constant it is advantageous to use a parametrization of the entropy in terms of temperature

$$U_i = c_i N_i R T \tag{5.15}$$

$$S_i = N_i s_{i0} + c_i N_i R \ln \left(\frac{T}{T_0} \right) + N_i R \ln \left(\frac{V}{V_0} \frac{N_{i0}}{N_i} \right) \tag{5.16}$$

We obtain an expression for the pressure (at constant temperature and thus constant internal energy)

$$\frac{P}{T} = \frac{\partial S}{\partial V} |_{U,N_i} = \sum_i N_i \frac{R}{V} = \sum_i \frac{P_i}{T} \tag{5.17}$$

where

$$P_i = N_i \frac{RT}{V} \tag{5.18}$$

are the **partial pressures** of the individual gases. These are the pressures that the individual gases exert on the walls.

Let us next introduce the total mole number $N = N_1 + N_2$ into the total entropy, also using the fact that the volume of the reference state is the same for both gases $v_0 = V_0/N_{i0}$

$$S = \sum_i \left[N_i s_{i0} + c_i N_i R \ln \left(\frac{T}{T_0} \right) + N_i R \ln \left(\frac{V}{N v_0} \right) \right] + \Delta S_{mixing} \tag{5.19}$$

The first term (in square brackets) is a collection of two gases in two different containers each at a density $N_1/V_1 = N_2/V_2 = N/V$. The last term

$$\Delta S_{mixing} = -R \sum_i N_i \ln \left(\frac{N_i}{N} \right) \tag{5.20}$$

is the entropy of mixing arising from the fact that pouring the two gases into the same container increases the entropy because we no longer have them separated. Indeed, $\Delta S_{mixing} > 0$ because $N_i/N < 1$ and thus $\ln N_i/N < 0$. Moreover, mixing requires a heat transfer of $Q_{mixing} = T\Delta S_{mixing}$. Because we performed mixing at constant temperature the internal energy does not change, implying that the heat transfer into the system must be compensated by work $W_{mixing} = -Q_{mixing}$. Mixing two gases is of course an irreversible process: once mixed it requires work to unmix them.

We now present an experiment that performs mixing and unmixing reversibly (see Figure 5.2). Insert into a vessel a double piston connected rigidly by external rods. Adjust the distance between the two pistons to be half the length of the vessel. Initially the vessel is divided into two halves by an impermeable wall. The left half of the vessel is filled with gas A only and the right half with gas B only. The total entropy is again the sum of the two. We arrange that the left piston at the right side of compartment A is permeable to gas A and we also make the right piston permeable to gas B. As we move the two rigidly connected pistons to the far left of the vessel, the two gases are mixed in the left compartment. As we move the pistons back to their original position de-mixing occurs. However, no work is done because both pistons only experience the constant partial pressure of gas B.

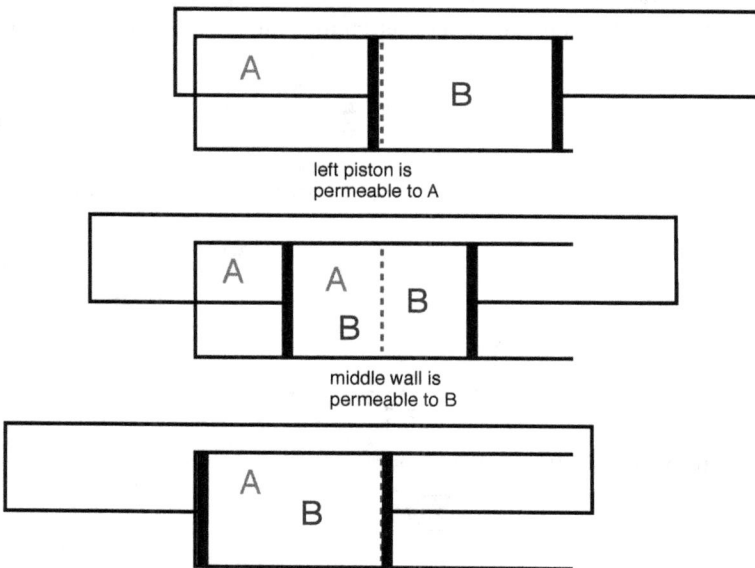

Fig. 5.2 A scheme to reversibly mix and unmix two gases. Initially two species (A and B) occupy separate chambers on the left and right hand sides of the apparatus respectively. As two rigidly connected pistons move to the left, species A is permitted to pass through the left piston (it is permeable to A but not to B). Meanwhile, B is allowed to pass through the dividing wall (which is permeable to B but not A). All other walls, including the rightmost piston, are impermeable to both A and B.

5.3   Gas reactions

The simplest reaction in a gas is the dissociation of molecules into atoms and the recombination of atoms into molecules. We write these reactions in a chemical way, here for the dissociation and formation of water

$$2H_2 + O_2 \rightleftarrows 2H_2O \tag{5.21}$$

or bringing all reactants to one side

$$0 \rightleftarrows -2H_2 - O_2 + 2H_2O$$
$$\rightleftarrows \sum_{i=1}^{3} \nu_i A_i \tag{5.22}$$

where A_i stands for H_2, O_2, and H_2O and $\nu_i = -2, -1, 2$ are stoichiometric coefficients. Now, chemical reactions change mole numbers so we should work with

$$dS = \sum_{i=1}^{r} \frac{\mu_i}{T} dN_i \tag{5.23}$$

Obviously the mole numbers change in proportion to the stoichiometric coefficients

$$dN_i = \nu_i d\widehat{N} \tag{5.24}$$

and we get

$$dS = \sum_{i=1}^{r} \frac{\mu_i}{T} \nu_i d\widehat{N} \tag{5.25}$$

In equilibrium we have $dS = 0$ or

$$\sum_i \nu_i \mu_i = 0 \tag{5.26}$$

For the dissociation of water this reads

$$2\mu_{H_2O} = \mu_{O_2} + 2\mu_{H_2} - E_{diss} \tag{5.27}$$

The dissociation energy E_{diss} accounts for the fact that to split a water molecule apart energy must be added to the system. Treating water vapor, oxygen, and hydrogen gases as ideal we get their chemical potentials directly from the entropy of a gas mixture by differentiation with respect to the mole numbers

$$\mu_i/RT = RT \left[\ln P + \ln x_i + \ln \phi_i(T)\right] \tag{5.28}$$

where we introduced mole fractions $x_i = N_i/N$; the functional form of $\phi_i(T)$ is irrelevant when we insist that the reaction occurs in an open dish at constant pressure and temperature. From our equilibrium condition we get

$$\sum_i \nu_i \mu_i = 0 \tag{5.29}$$

$$\sum_i \nu_i \ln x_i = -\sum_i \nu_i \ln P - \sum_i \nu_i \phi_i(T) - E_{diss}/RT \tag{5.30}$$

or

$$\prod_i x_i^{\nu_i} = P^{-\Sigma_i \nu_i} K(T) \tag{5.31}$$

This is called the mass action law and

$$K(T) = \exp[-\Sigma_i \nu_i \phi_i(T) - E_{diss}/RT] \tag{5.32}$$

is the equilibrium constant. The latter is listed in tabular form for many chemical reactions. A chemical reaction never proceeds to the absolute minimum in energy as one would expect from a mechanical point of view (equivalent to thermodynamics at absolute zero). Rather, the extent of a chemical reaction at finite temperature is governed by the principle of maximum entropy which leads to the condition that the chemical potential of the reaction species must be equal.

The mass action law is just one equation to determine the equilibrium concentrations (or mole fractions) in a reacting gas mixture, whereas in water dissociation there are three distinct species. We thus need, not surprisingly, another two equations. Obviously the final distribution of water, oxygen, and hydrogen molecules depends on how much material we put into the reaction chamber, for instance the mole numbers of oxygen and hydrogen. If we start with only water vapor, then dissociation will produce twice as much hydrogen as oxygen. Likewise, if initially we start with an abundance of hydrogen and small amount of oxygen, then at the end of the reaction we have a hydrogen-rich gas.

For a simpler system, namely the dissociation equilibrium for hydrogen alone, $H_2 \rightleftarrows 2H$, the mass action law can be written as

$$\frac{x^2}{1-x} = \frac{g_a^2}{g_m} (\pi m_a)^{3/2} \frac{(k_B T)^{5/2}}{h^3 P} \frac{1 - \exp[-\Theta_v/T]}{T/2\Theta_r} \exp[-E_{diss}/k_B T] \tag{5.33}$$

Here we introduced a few new quantities such as the mass of a hydrogen atom, m_a as well as the degeneracies of the atomic and molecular ground

Thermodynamics

states, g_a and g_m. Θ_v and Θ_r are the vibrational and rotational temperatures of the hydrogen molecule. Vibrations and rotations constitute additional degrees of freedom in which energy is stored. In Figure 5.3, we show the mole fraction of atoms as a function of temperature for various pressures.

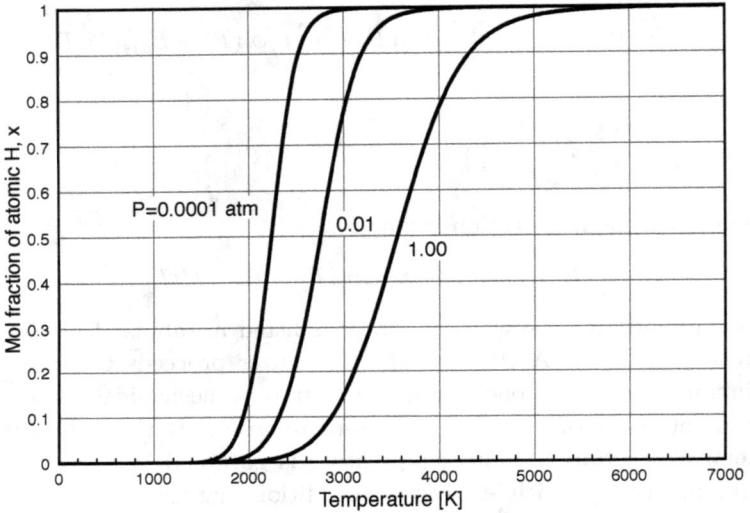

Fig. 5.3 The mole fraction of atomic hydrogen in a gas at various pressures as a function of temperature. The parameters for a hydrogen molecule are: $m_a = 1.67 \times 10^{-27}$ kg, $g_a = 2, g_m = 1$, $\Theta_r = 85.4$ K, $\Theta_v = 6100$ K, and $E_{diss} = 4.53$ eV.

The exponential in the mass action law is an indication that chemical reactions are thermally driven processes with high temperatures needed to overcome an activation barrier, in this case the dissociation energy. For H_2, the dissociation energy, E_{diss}, corresponds to a temperature of $T_{diss} = E_{diss}/k_B T = 52,200$ K. One could naively argue that the atomic mole fraction should only become significant if the gas is at about that temperature. However, such an argument is always false as the pre-factor before the exponential is almost never of order one. Indeed, for hydrogen dissociation it is so large that atomic hydrogen already appears around 2500 K. By 6000 K, see Figure 5.3, the system is essentially atomic. It is by the way, absolutely essential for life on earth as we know it that the atomic fractions of the major gases in the earth's atmosphere, hydrogen, oxygen, and nitrogen, are negligible, as they are chemically very reactive and destructive to most life forms.

5.4 Blackbody radiation

We have already discussed the fact that the phenomenon of blackbody radiation makes for a perfect thermometer to measure the temperature of remote objects. Recall that a blackbody is an empty cavity of any shape and of any material, empty except for electromagnetic radiation in equilibrium at some temperature. Such radiation is pervasive. If there is a gas present in the cavity, it will eventually thermalize with the blackbody radiation by radiative energy transfer. In an empty cavity surrounded by walls you must imagine black body radiation as standing electromagnetic waves with a distribution of wavelengths according the Planck radiation law. Because of the wave-particle dualism in quantum mechanics we can also view blackbody radiation as a collection of photons of different energies $h\nu$. Blackbody radiation is an example to demonstrate that thermodynamics can deal with any system in equilibrium and is not restricted to gases. Indeed, it is amazing that for blackbody radiation we obtain the perfect fundamental relation (valid for all pressures and temperatures) which, if you recall, was not the case for an ideal gas at all! (There were of course well-explained reasons for this, see the discussion around equation (4.28)).

We now want to look in detail at the thermodynamics of a blackbody radiator. As input we need results from measurements of the equations of state as summarized in the Stefan-Boltzmann law

$$U = bVT^4 \qquad (5.34)$$

$$P = \frac{U}{3V} \qquad (5.35)$$

where $b = 7.56 \times 10^{-16}$ J/m^3 K^4. With the discovery of his blackbody radiation law Planck was able to express this constant in terms of fundamental constants as $b = \pi^2 k_B^4/(15\hbar^3 c^3)$. How can you measure these two relations in a lab experiment involving a box surrounded by walls? Punching a little hole into the box serves as the source of radiation emanating from the box. Likewise, radiation striking the hole and entering the cavity will scatter off the interior walls many times with the chance of exiting through the small hole being minimal and is given by the ratio of the area of the hole to the total area of the inside walls. This also means that any light exiting the hole must be in thermal equilibrium inside the cavity. Note that the mechanism by which photons exchange energy with the walls is unimportant to the thermodynamics!

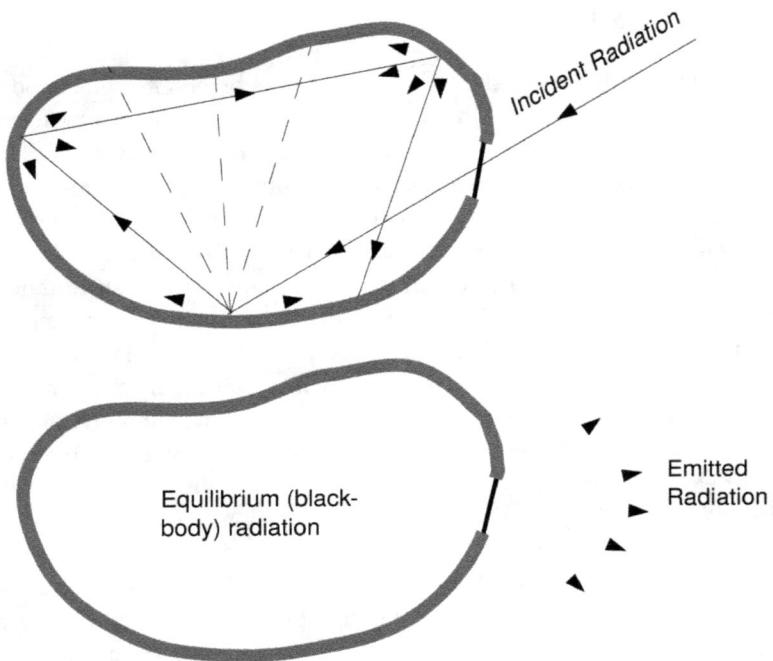

Fig. 5.4 A small hole in a hot cavity makes for a perfect blackbody radiator. Incident light that enters through the opening is internally reflected many times before it is allowed to escape. This results in thermal equilibration. The wavelength of the emitted light is therefore fully determined by the temperature of the cavity.

The temperature can be measured by a thermometer stuck into the box. The radiation can be used to heat up water at a rate dU/dt over a sufficiently long time period. The radiation pressure is best understood in the photon picture because each photon has a momentum $h\nu/c$ and will transfer twice that amount to a surface from which it is reflected. The pressure is then given by twice the ratio of the radiation strength and the speed of light. As an example, the radiation strength of the sun at the surface of the earth is 1.4 W/m^2, so that the sun's radiation pressure on earth is about 10^{-5} Pa $= 10^{-10}$ atm. Even compared to the radiation pressure of sound waves, it is a factor 10^{-6} smaller at the same radiation strength.

Note that no mole number appears in these equations because the number of standing waves, or photons, in the cavity changes as temperature changes and thus is not a fixed quantity and the corresponding chemical potential is zero. Because U and V occur, but not S, we will work in the entropy representation for which we get

$$\frac{1}{T} = b^{1/4} \left(\frac{V}{U}\right)^{1/4} \tag{5.36}$$

$$\frac{P}{T} = \frac{1}{3}b^{1/4} \left(\frac{U}{V}\right)^{3/4} \tag{5.37}$$

Inserted into the Euler relation, we get the entropy of blackbody radiation

$$\begin{aligned} S &= \frac{1}{T}U + \frac{P}{T}V \\ &= \frac{4}{3}b^{1/4}U^{3/4}V^{1/4} \end{aligned} \tag{5.38}$$

5.5 Polymers

Polymers are long chain molecules in which the same subunits, called monomers, are repeated many times. Examples are alkanes $CH_3-(CH_2)_n-CH_3$, poly(ethanolglycol) or PEG, $CH_3O - (CH_2CH_2O)_n - CH_3$; DNA, etc. The $n = 1$ alkane is propane; $n = 8$ is octane, etc. PEG is used in cosmetics, medical applications such as laxatives and much more.

Along the polymer the monomers can be oriented in different directions with respect to their neighbors. For instance, for $C - C$ and $C - O$ bonds, a third bond along the backbone of the polymer can be in three different directions with respect to the plane formed by the two previous bonds, namely in the same plane (that is called a *trans* configuration) or at about $70°$ above or below the plane, called *gauche* configurations.

From this degree of freedom alone, a polymer consisting of n monomers already can exist in 3^n distinct conformations. This leads to the inherent complexity of polymers (see Figure 5.5).

To understand the macroscopic properties of single polymers and polymer solutions such as paints and glues one needs to measure their thermal and mechanical equations of state. This seems tricky to do for single polymers but has been achieved in the mid 1990's using the Atomic Force Microscope (AFM) or optical tweezers. The principle of operation of the AFM is to chemically bind one end of a polymer to a substrate and the other end to a cantilever and move the cantilever to apply a force, see Figure 5.6.

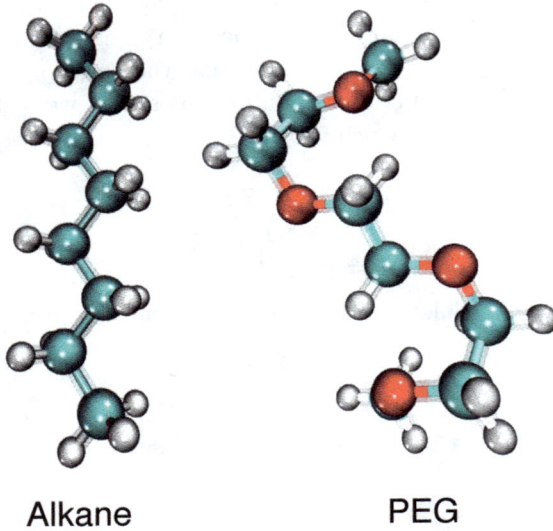

Alkane   PEG

Fig. 5.5   An alkane polymer with six monomers and a PEG polymer with three monomers.

The far end of the cantilever is set at a distance D from the substrate via a piezoelectric actuator, the deflection of the cantilever is then measured optically by deflection of light from the back of its tip. This information can be used to determine both the length of the polymer and the applied force (via the cantilever spring constant). Optical tweezers are used in the same manner as an AFM with the exception that the force is applied via an optical trap rather than a Hookean spring, see Figure 5.7. In an optical trap, radiation pressure from the incident laser light causes forces on the surface of a refractive sphere; the gradient of the laser intensity is such that the forces always point to the center of the trap. A typical force-extension curve is shown in Figure 5.8. Note that a polymer is a one-dimensional object so that the extensive variable V goes over into length L and the force f replaces the pressure. The length here is the end-to-end length between the first and the last C atoms in the chain. Another length in a polymer is the contour length L_c which is the length along the backbone.

Also plotted in Figure 5.8 is a fit with a simple model, the Wormlike Chain

$$f = \frac{k_B T}{L_p(T)} \left[\frac{L_c}{4(1 - L/L_c)^2} - \frac{1}{4} + \frac{L}{L_c} \right] \tag{5.39}$$

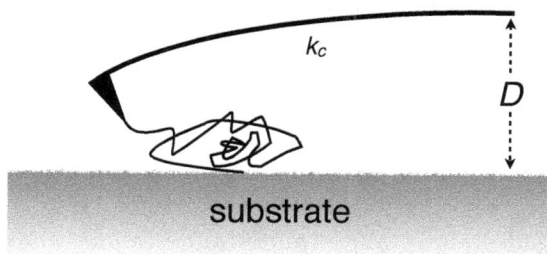

Fig. 5.6 Schematic representation of an AFM experiment with a harmonic cantilever, spring constant k_c, at a distance D above the surface.

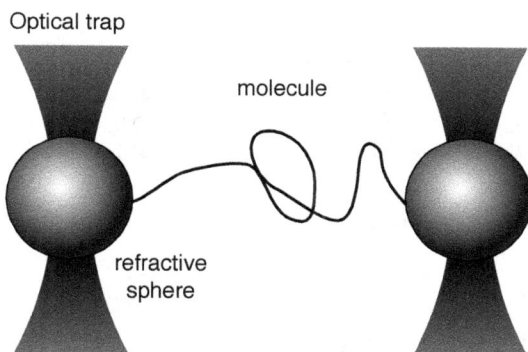

Fig. 5.7 Schematic representation of an experiment using optical tweezers. The case where a molecule is bound between two optical traps is shown; it is also possible to make use of a single trap and bind the other end of the molecule to a substrate.

Here L_p is called the persistence length which is a measure for the distance along the chain over which there is a correlation in orientation, i.e. before the chain makes, on average, a drastic bend. For flexible chains, L_p is typically a few monomer lengths. Considering that the WLC formula has only one fitting parameter, L_p, the fit is pretty good. It again makes the point that very complicated systems such as DNA or a polypeptide can have simple macroscopic properties worth examining with thermodynamics.

All we have from this stretching experiment is the mechanical equation of state f/T as a function of T and L but we have no measurement for the thermal equation of state. Thermodynamics can help to a certain extent.

Remember that $f/T = (\partial S/\partial L)|_U$ and $1/T = (\partial S/\partial U)|_L$. Also recall

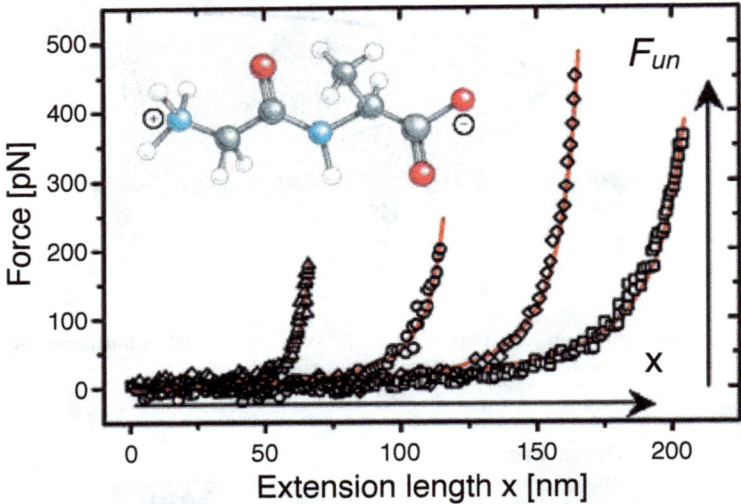

Fig. 5.8   Stretching a polypeptide polymer with an AFM, shown for different lengths, the longest consisting of about 200 monomers. The fit is the Wormlike Chain.

that second derivatives do not depend on the order of differentiation, i.e.

$$\frac{\partial^2 S}{\partial L \partial U} = \frac{\partial^2 S}{\partial U \partial L} \tag{5.40}$$

or

$$\frac{\partial}{\partial L} \frac{1}{T}\Big|_U = \frac{\partial}{\partial U} \frac{f}{T}\Big|_L \tag{5.41}$$

Let us write the WLC formula as

$$\frac{f}{T} = \frac{1}{L_p(T)} \tilde{f}(L) \tag{5.42}$$

to get

$$\frac{1}{\tilde{f}(L)} \frac{\partial}{\partial L} \frac{1}{T}\Big|_U = \frac{\partial}{\partial U} \frac{1}{L_p(T)}\Big|_L \tag{5.43}$$

This is as far as thermodynamics can get us. To proceed we need information on the dependence of the persistence length on temperature. Lacking this information we might argue that the persistence length will decrease as a function of temperature $L_p(T) = L_p^0 T_0/T$ simply because at higher temperature the chain is more flexible. In this case we can rewrite the above equation as

$$\frac{\partial}{\partial g}\frac{1}{T}\Big|_U = \frac{\partial}{\partial U}\frac{1}{T}\Big|_L \tag{5.44}$$

$$g = \frac{1}{L_p^0}\int_0^L \tilde{f}(L)dL' \tag{5.45}$$

and see that $1/T$ must be a symmetric function of U and g. If we require that the internal energy increases with temperature as in an ideal gas the simplest relation would be

$$\frac{1}{T} = \frac{cL_p^0}{U + g(L)} \tag{5.46}$$

$$U = cL_p^0 T - g(L) \tag{5.47}$$

Thus we have obtained the entropy for a hypothetical extension of the WLC model to be

$$S = \frac{1}{T}U + \frac{f}{T}L \tag{5.48}$$

What thermodynamics thus tells us is that the internal energy as a function of T and L must be either measured or calculated. Statistical mechanics has indeed done that job.

5.6 Thermodynamics of adsorbates

Wolfgang Pauli (1900-1958)
"The surface was invented by the devil."

Because macroscopic objects are all finite they must terminate in an interface or surface to another medium. As an example a liquid surface is actually the interface with its own vapor pressure: if we were to pour a liquid into a closed container that has been completely pumped out it would start to evaporate until the vapor pressure above the liquid is such that the rate of evaporation equals the rate of condensation thus establishing equilibrium. The (saturated) vapor pressure of water at room temperature is 25 millibar=2.5 kPa or about 2.5% of atmospheric pressure.[1]

What about the surface of a solid in a vacuum chamber? It will also evaporate or sublimate. However, the vapor pressure of most solids, such as

---

[1]Relative humidity is the ratio of the actual vapor pressure to the saturated vapor pressure. The boiling point of water is at a temperature where its partial vapor pressure equals that of the atmosphere above it.

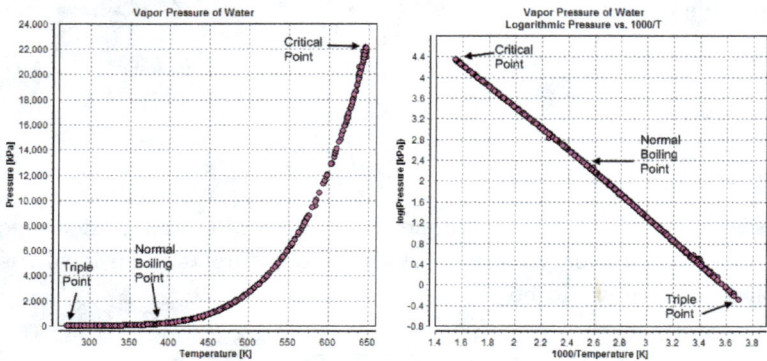

Fig. 5.9   The saturation vapor pressure of water as a function of temperature. Image adapted from [Cordes (2009)].

metals below their melting point, is usually minute. For instance for gold it is 2.37×10^{-4} Pa or 10^{-9} atm at its melting temperature of 1337.73 K = 1064.58° C. For harder solids it is even less. Also the rate of sublimation for solids is minute except at very high temperatures. So we can treat the surface of a solid as almost "inert"? Does this imply that nothing happens at solid surfaces when exposed to the atmosphere? No! Indeed, a lot happens, some good and some bad! Iron surfaces rust forming iron oxides of all sorts, silver surfaces turn black with silver sulfides when exposed to small amounts of hydrogen sulfide in the air, copper surfaces turn green with a variety of oxides, sulfides, and chlorides, solid salt adsorbs water from the air and turns mushy.

Of the chemical elements only gold has an inert surface that does not change over time. That, by the way, is the reason why gold has been admired throughout historical times, not because it is rare (it is more abundant than platinum, osmium, iridium, rhodium, ruthenium, etc.).

What then is the good side of surfaces? Because surfaces can adsorb gas molecules which subsequently react, one can use specific surfaces (natural or designed) to trigger reactions that will not happen in the atmosphere. Such surfaces are called catalysts. Fractionation of crude oil into gasoline and other products is done over catalysts, as is the production of most polymers and plastic products. In the Haber-Bosch reaction atmospheric nitrogen molecules, N_2, are dissociated over a very special iron ore catalyst into atomic nitrogen and then combined with dissociated hydrogen to form ammonia NH_3. As a fertilizer, ammonia has heralded in the Green Revolu-

Fig. 5.10 The Old City Hall of Hamburg with its beautiful copper roof.

tion, increasing food production on earth a thousandfold. Like everything good invented in science, the production of ammonia had its negative impact because it is also used as the starting material of TNT: Trinitrotoluene and nitroglycerin. It is estimated that the First World War was prolonged by at least two years because of the sudden availability of ammonia through the Haber-Bosch process.

Let us return to thermodynamics. If the interaction between gas molecules and the surface of a solid is attractive, the molecules will be trapped in the surface region and form an adsorbate with a density greatly enhanced relative to that of the gas far from the surface, see Figure 5.12. Let us divide the volume in front of the solid surface into an outer region V_g associated with the gas phase and a surface region V_s of thickness z_s for the adsorbate.

$$V = V_g + V_s \qquad (5.49)$$

We keep z_s arbitrary for the moment. Likewise, we split the the total amount of gas, N moles, into gas phase and surface contributions

$$N = N_g + N_s \qquad (5.50)$$

We should not attempt to identify N_s as the actual amount of gas molecules in the surface region because V_s is unknown and will of course

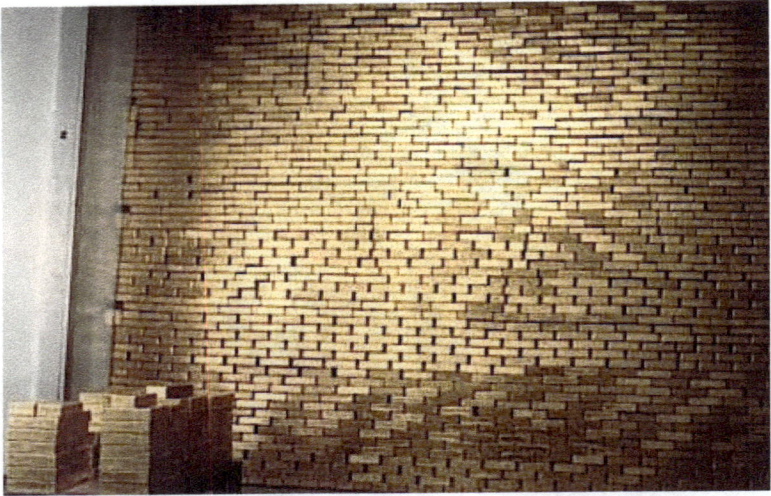

Fig. 5.11   A stack of gold bars in the vault of the New York Federal Reserve. Each gold bar weighs about 27.4 pounds and was worth about \$350,000 in 2009.

depend on the seizure point z_s. Rather we define

$$N_g = V_g \rho_g / M \tag{5.51}$$

in terms of the bulk mass density ρ_g and the molar mass M of the gas. N_g is then the amount of gas that would occupy the volume V_g if the local gas density $\rho(z, \mathbf{R})$ were the same everywhere, which, of course, it is not (it increases towards the surface). $N_s = N - N_g$ is then the surface **excess** mole number in the surface region V_s.

Similar excess quantities are defined for energy and entropy by writing

$$U = U_g + U_s$$
$$S = S_g + S_s$$
$$U_g = u_g V_g \rho_g / M$$
$$N_g = s_g V_g \rho_g / M \tag{5.52}$$

so that $u_g = U_g / N_g$ and $s_g = S_g / N_g$ are the energy and entropy per mole far from any surface. Surface excess energy and entropy follow straightforwardly.

Because all surface excess quantities are defined as differences, their new values after a change in the choice of z_s can easily be calculated: assume that changing z_s alters V_s by ΔV_s and thus V_g by $-\Delta V_g$ because V is fixed.

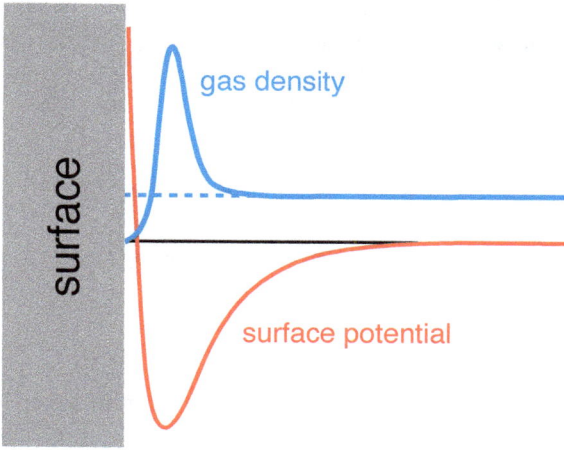

Fig. 5.12   The density enhancement in front of the surface constitutes the adsorbate: it results from the attractive force/potential that the solid exerts on gas particles. The dashed line illustrates the Gibbs convention $V_s = 0$.

Then the excess quantities change like

$$\Delta N_s = \Delta V_s \rho_g / M$$
$$\Delta U_s = \Delta V_s u_g \rho_g / M$$
$$\Delta S_s = \Delta V_s s_s \rho_g / M \tag{5.53}$$

which can be calculated knowing bulk values only, i.e. without needing any microscopic information, as befits a good thermodynamic theory.

What value of z_s one chooses in a particular situation is a matter of convenience. Most frequently, the Gibbs convention $V_s = 0$ is advantageous. Suffice it to say that any quantity of physical significance in surface thermodynamics cannot and will not depend on such a convention.

Our next task is to include surface effects into the fundamental relation. For the gas phase component we have in the energy representation

$$dU_g = TdS_g - PdV_g + \sum_{k=1}^{c} \mu_k dN_g^{(k)} \tag{5.54}$$

$$T = \frac{\partial U_g}{\partial S_g}\Big|_{V_g, N_g} \tag{5.55}$$

and so on. Of course, the Euler equation and the Gibbs-Duhem relation is also valid for this phase.

With the adsorbate present, the total energy must be a function of an additional extensive variable, which must be the surface area, A, of the solid

$$U = U(S, V, A, N_1, ..., N_c)$$

$$dU = TdS - PdV + \sum_{k=1}^{c} \mu_k dN^{(k)} + \gamma dA \tag{5.56}$$

where we introduced the intensive variable conjugate to the area as the surface tension

$$\gamma = \frac{\partial U}{\partial A}\Big|_{S,V,N^{(k)}} \tag{5.57}$$

Its significance can be seen in the following experiment: let one of the walls of the volume V, assumed for simplicity to be a cube of side length L, be a piston. Moving it adiabatically by a distance dL by applying a force f requires an energy

$$dU = -fdL \tag{5.58}$$

In bulk thermodynamics one argues that the external force balances the pressure exerted by the molecules inside, i.e. $F = PL^2$, in order for dU to remain small in a quasi-static process. Doing this identification one ignores effects at the edges of the piston. For the adsorbate, however, those edge effects are important because it is at the edges where the surface area A is increased. Indeed, surface tension will act against the motion of the piston so that the net force required for a change dU is

$$fdL = PL^2 dL - (4L)\gamma dL$$
$$= PdV - \gamma dA \tag{5.59}$$

and the change in internal energy becomes

$$dU = -PdV + \gamma dL \tag{5.60}$$

Let us next deal with the excess quantities and recall that the solid provides ideal walls, i.e. mechanically and thermally inert, so that

$$dN^{(k)} = dN_g^{(k)} + dN_s^{(k)} = 0$$
$$dS = dS_g + dS_s = 0 \tag{5.61}$$

We then get from (5.60)

$$dU = dU_g + dU_s = -PdV_g - PdV_s + \gamma dA \tag{5.62}$$

Using (5.54) we can solve for

$$dU_s = TdS_s - PdV_s + \sum_k \mu_k dN_s^{(k)} - \phi dA \tag{5.63}$$

where

$$\phi = \gamma_0 - \gamma = -\frac{\partial U_s}{\partial A}\Big|_{S_s, V_s, N_s^{(k)}} \tag{5.64}$$

is the spreading pressure of the adsorbate, with γ_0 as the tension of the clean surface without the adsorbate. The spreading pressure (dimensionally a force) contains the quasi two-dimensional adsorbate within the area A. If the adsorbate is dilute, it satisfies a two-dimensional ideal gas law and we have

$$\phi = \frac{N_s}{A}RT \tag{5.65}$$

The Gibbs-Duhem equation for a one-component adsorbate reads

$$S_s dT - V_s dP + N_s d\mu - A d\phi = 0 \tag{5.66}$$

According to this equation the chemical potential is a function of three variables $\mu = \mu(T, P, \phi)$ which is in contradiction to our general statement that a one-component system has only two degrees of freedom. The elegant way out of this apparent contradiction is to eliminate either the dP or the $d\mu$ term using the Gibbs-Duhem relation for the bulk. An easier way is to make use of the arbitrariness of the adsorbate volume by setting $V_s = 0$. On the other hand, eliminating the chemical potential leads to a useful result

$$\frac{A}{N_s}d\phi + \frac{V_s}{N_s}dP - \frac{S_s}{N_s}dT = \frac{V_g}{N_g}dP - \frac{S_g}{N_g}dT \tag{5.67}$$

or

$$\Gamma^{-1}d\phi = \left(\frac{V_g}{N_g} - \frac{V_s}{N_s}\right)dP - \left(\frac{S_g}{N_g} - \frac{S_s}{N_s}\right)dT \tag{5.68}$$

connecting adsorbate properties to those of the bulk gas phase. This is the Clausius-Clapeyron equation for adsorbates. Here we defined the adsorbate density or surface coverage by

$$\Gamma = \frac{N_s}{A} \tag{5.69}$$

To see its usefulness let us integrate it under isothermal conditions to get

$$\phi(T, P) = \phi_0(T, P_0) + \int_{P_0}^{P} \Gamma(T, P)\left(\frac{V_g}{N_g} - \frac{V_s}{N_s}\right)dP \tag{5.70}$$

To fix the reference point judiciously, we note that at zero pressure there is no gas and thus no adsorbate. This gives

$$\phi(T,P) = \int_0^P \Gamma(T,P) \left(\frac{V_g}{N_g} - \frac{V_s}{N_s} \right) dP \tag{5.71}$$

Again we choose $V_s = 0$ and use the ideal gas law for the bulk to get

$$\phi(T,P) = RT \int_{P_0}^P \Gamma(T,P) \frac{1}{P} dP \tag{5.72}$$

which allows us to determine the spreading pressure from measurements of the surface coverage.

Example 5.1. It is found experimentally that for many systems only a single layer of molecules, a monolayer, can be adsorbed on a solid surface such as metals. The reason for this is that gas molecules, such as carbon monoxide, bind to metal atoms of the surface forming a chemical bond. A crude model to describe such systems is to assume that the adsorbate molecules do not interact with each other; this is oversimplified because in a monolayer the adsorbate molecules are within a few Angstroms of each other and interact strongly. Ignoring this knowledge we need at constant temperature a function $\Gamma(P)$ that is zero at zero pressure and approaches a maximum Γ_{max} at high pressures. The simplest such function is the Langmuir isotherm

$$\theta(P,T) = \frac{\Gamma(P,T)}{\Gamma_{max}} = \frac{P}{P + P_0(T)} \tag{5.73}$$

where θ is called the surface coverage. Inserted into (5.72) we then get

$$\phi(T,P) = RT\Gamma_{max} \ln(P/P_0 + 1) \tag{5.74}$$

If one adsorbate molecule adsorbs per surface atom, the maximum surface concentration is of the order of $10^{19}/m^2$ or $10^{15}/cm^2$.

Remark 5.2. Let us return to the example discussed at the beginning of Chapter 3 with respect to the interpretation of the entropy as a measure of disorder. Let us identify the "objects" as adsorbing molecules; there are N_A of them that can be put on N_{site} adsorption sites for a total of $\Omega = \binom{N_{site}}{N_A}$ choices and the entropy is $S = k_b \ln \Omega$. Using Stirling's formula for large numbers

$$\ln N! \approx N \ln N - N \tag{5.75}$$

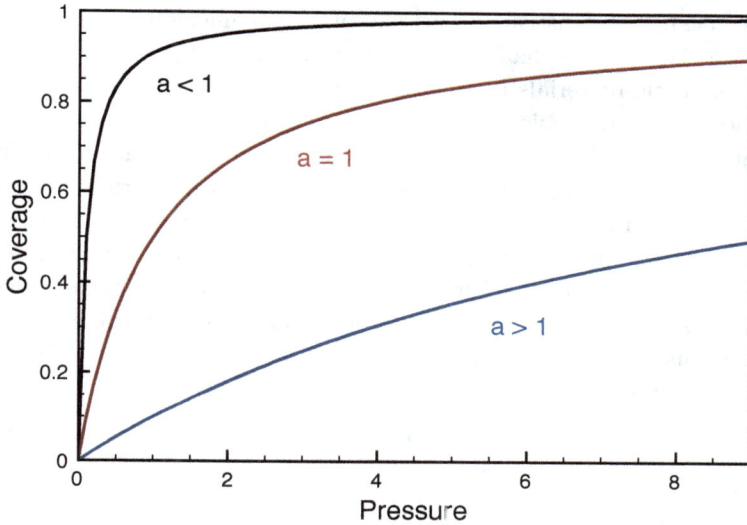

Fig. 5.13   Langmuir isotherms: the coverage, normalized to one with the maximum coverage, as a function of pressure for parameters $a = P_0(T)$.

the entropy as a function of the relative coverage $\theta = N_A/N_{site}$ reads

$$S/k_b = -N_{sites}\left[(1 - \theta)\ln(1 - \theta) + \theta\ln\theta\right] \qquad (5.76)$$

The entropy is zero for a clean surface ($\theta = 0$) and for a completely covered surface ($\theta = 1$), and disorder is maximal when half the surface is covered.

5.7   Magnetic systems

In this section we will show how to incorporate magnetic phenomena into thermodynamics dealing with diamagnetism, paramagnetism, and ferromagnetism.

5.7.1   *Diamagnetism*

We first consider non-magnetic materials in which the constituent atoms or molecules carry no permanent magnetic moments. Because electrons are spin one-half particles with magnetic dipoles, non-magnetic materials must be made up of even numbers of electrons (called closed-shell atoms or molecules). The constituent particles of atomic nuclei, protons and neutrons, are also spin one-half particles with magnetic dipoles, so the nuclei of non-magnetic materials must also be made up of even numbers of protons and neutrons. Noble gases and homo-nuclear dimers, such as O_2, in their quantum mechanical ground state are examples. Electrons orbiting the nuclei can be regarded as current loops that generate intrinsic magnetic fields which are randomly oriented in the absence of an external magnetic field. Thus they are non-magnetic. Placing such a system in an external magnetic field of intensity H will orient these atomic/molecular currents, inducing a secondary macroscopic magnetic field that, according to Lenz's rule, weakens the primary applied field. Consequently, the total magnetic induction (or flux density) is

$$\mathbf{B} = \mathbf{B}_0 + \mathbf{B}' \tag{5.77}$$

$$= \mu_0(\mathbf{H} + \mathbf{M}) \tag{5.78}$$

where $\mu_0 = 4\pi \times 10^{-7} Wb/(A\ m)$ (Weber per Ampere and meter) is the magnetic susceptibility of empty space and $\mathbf{M}$ is the magnetic moment (or magnetization) of the system. Because the dipole moment is induced by the external field and is typically small, it must be proportional to the applied field, i.e.

$$\mathbf{M} = \chi \mathbf{H} \tag{5.79}$$

Here χ is the magnetic susceptibility of the material. In most materials this diamagnetism is negligible. Some exceptions are alkali halides (for example NaCl, LiBr) and similar solids where diamagnetism is noticeable. A more important group of materials are superconductors in which, according to the Bardeen-Cooper-Schrieffer (BCS) theory, conduction electrons are paired to zero-spin Cooper pairs and diamagnetism is so large as to cancel any applied field. This expulsion of an applied field is the so-called Meissner effect, one of the most stunning properties of superconductors.

Remark 5.3. The relation (5.79) is only valid for isotropic materials for which χ is a scalar. For anisotropic materials χ is a tensor. The diamagnetic susceptibility can be calculated quantum-mechanically in first-order perturbation theory which gives

$$\chi = -\frac{Ze^2 \widetilde{N}}{6mc^2} \overline{r^2} \tag{5.80}$$

where Z is the number of electrons in each of the $\widetilde{N}$ atoms (or molecules) in the sample. The constants e and m are the charge and mass of an electron respectively, while c is the speed of light, and $\overline{r^2}$ is the average mean square radius of the electron cloud. If the atoms or molecules have low-lying excited states there is an additional contribution to the diamagnetic susceptibility called *van Vleck* diamagnetism. In a metal the gap between the ground state and the excited states closes and a conduction band forms that is filled up to the Fermi energy $E_F = \hbar^2 k_F^2 / 2m$. The quantum mechanical theory becomes rather complicated resulting in Landau diamagnetism for which the susceptibility is also constant

$$\chi_L = -\frac{e^2 k_F}{12\pi^2 mc^2} \qquad (5.81)$$

5.7.2 *Paramagnetism*

The degeneracy of two electrons in the same orbital state is lifted by a magnetic field, leading to the Zeeman effect: The energy of the electrons with the spin parallel to the field is lowered by $\mu_e H$ and raised for those antiparallel to the field. This leads to a macroscopic magnetic moment called Pauli paramagnetism. In a metal, the Pauli paramagnetic susceptibility is (for a free electron metal) three times as large as the Landau diamagnetic susceptibility (5.81) (but of opposite sign).

Let us next examine a material where the constituent atoms have an odd number of electrons. This implies that each atom carries a spin and thus a magnetic dipole $\boldsymbol{\mu}$. In the absence of an external field these dipoles are oriented randomly. However, in a magnetic field orientation takes place resulting in an overall magnetic moment that, for small fields, is proportional to the applied field with a susceptibility

$$\chi_{Curie} = \frac{1}{3} \frac{\widetilde{N} < \mu^2 >}{k_B T} \qquad (5.82)$$

This is called Curie's law of paramagnetism. It is well satisfied for solids in which the magnetic atoms or ions are sufficiently separated so that their magnetic interaction is negligible.

In many systems, however, the magnetic atoms are so close together that their magnetic interactions cannot be neglected. To account for this Pierre Weiss introduced a mean field

$$\mathbf{H}_I = \lambda \widetilde{N} < \boldsymbol{\mu} > \qquad (5.83)$$

as acting on average on any one of the magnetic atoms as the result of its interactions with all others. The total field on such an atom is this Weiss field plus the externally applied field. It is again beyond the scope of thermodynamics to derive the expression for the susceptibility, but we quote the result

$$\chi_{CW} = \frac{\widetilde{N}\mu^2}{3k_B(T - T_C)}$$

$$T_C = \lambda \frac{\widetilde{N}\mu^2}{3k_B} \qquad (5.84)$$

This is the Curie-Weiss law which describes the paramagnetism in many solids quite well. Because the strength parameter in the mean field is undetermined one takes the Curie temperature T_C as a characteristic property obtained from experiment.

We note that the Curie-Weiss susceptibility has a singularity at the Curie temperature. For $T > T_C$ the system is paramagnetic. However, for $T < T_C$ there is spontaneous magnetization even in the absence of a field which reaches a maximum at $T = 0$. The system is ferromagnetic.

5.7.3 *Ferromagnetism*

To understand ferromagnetism let us take a piece of iron and heat it above its Curie temperature $T_C = 1043$ K to make it paramagnetic. We then cool it down to room temperature in the absence of a magnetic field: the iron does not show any magnetic moment. As we apply an increasing magnetic field, a magnetic moment is measured that, for large enough fields, saturates (see the inner curve on the left of Figure 5.14).

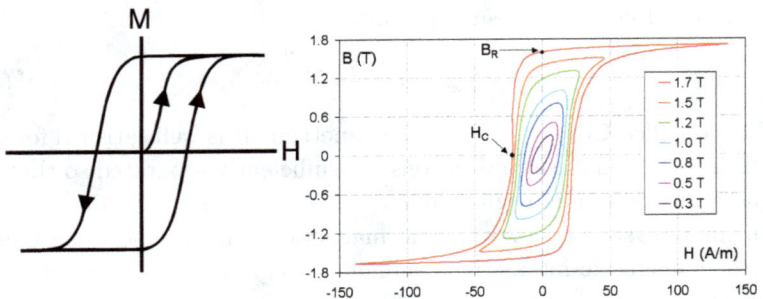

Fig. 5.14 Hysteresis in the magnetization of a ferromagnet. Left: inner curve: magnetization of a demagnetized specimen. Right: magnetization loops for different maximum applied fields in Tesla. Image adapted from [Zureks (2006)].

We then reduce the magnetic field gradually and trace out the upper curve in the left image of Figure 5.14. As the field is turned off some remnant magnetization remains: we have magnetized the iron and created a permanent magnet. As we reverse the magnetic field the magnetic moment is further reduced, eventually to zero, and then attains the opposite sign until saturation is reached. If we then reduce this field and eventually reverse the polarity again, the lower curve in the left image of Figure 5.14 is followed. The whole magnetization curve is called a hysteresis loop. It shows that the magnetization at any given external magnetic field is dependent on the history of how this field was applied: the system is NOT in thermodynamic equilibrium. Moreover, magnetizing a ferromagnetic material is NOT reversible. This fact is even more pronounced if we go around a hysteresis loop to a smaller maximum field strength: we get hysteresis loops that are nestled inside the one that reaches saturation magnetization (see the right panel in Figure 5.14). The width of the hysteresis loop at zero magnetization is called its coercivity. It is large for hard magnetic materials such as permanent magnets and the magnetic storage bits in hard disks. Soft magnetic materials such as the Heusler alloys (Cu_2MnSn, Ni_2MnAl, and others) have small coercivity.

Remark 5.4. The physical origin of ferromagnetism: According to classical electromagnetism, two nearby magnetic dipoles will tend to align in opposite directions. In ferromagnetic materials, however, this ordering is precluded by the quantum mechanical exchange interaction. To understand this we develop a simple thought experiment: take two identical atoms, each with a single electron in its highest occupied orbital. As long as these two atoms are far apart they do not interact and the spin orientations are random. As they are moved closer together their orbital wave functions overlap and the Pauli exclusion principle states that the total wave function for both electrons must be antisymmetric. This can be achieved in two ways: (1) they remain in the same orbital, in which case their spins must align antiparallel, or (2) they arrange their orbital wave function to be antisymmetric, in which case their spins must be parallel. In ferromagnetic materials the latter scenario has the lower energy. In antiferromagnetic materials the former scenario wins. In iron the exchange interaction is about a factor 10^3 times stronger than the dipole-dipole interaction.

This energy argument would suggest that all ferromagnets are always maximally magnetized, which they are not. Indeed, a good thermodynamicist knows that energy alone is not sufficient. With all spins aligned parallel to each other, the spin degrees of freedom would be perfectly ordered, i.e.

the contribution to the entropy would be zero. However, thermodynamics says that this can only happen at absolute zero. Thus nature must introduce some spin disorder. This happens as follows: in a ferromagnet small regions, called magnetic or Weiss domains, are ordered. If the whole sample is unmagnetized these Weiss domains are randomly oriented and are separated from each other by spatially narrow and disordered regions or Bloch walls. Thus creating magnetic order in the Weiss domains is energetically favorable. But to increase the entropy these domains must be randomly oriented and separated by disordered Bloch walls. As we apply a magnetic field the Weiss domains re-arrange in the field direction to create a net magnetization. Removing the field some of these re-arranged domains remain oriented because there is an energy barrier to overcome to reach the state of total disorder again. This struggle between favoring a lower energy and at the same time struggling to maximize the entropy results in hysteresis. Once the temperature is raised too high, i.e. above the Curie temperature, thermal motion becomes dominant and disorder among the domains is restored: the material becomes paramagnetic.

Remark 5.5. The transition between the ferromagnetic and paramagnetic state is an example of a phase transition which we will study in Chapter 10. There exists a simple model called the Ising spin model that serves as the basis for a statistical mechanical theory of critical phenomena and calculates the details of the transition.

5.7.4   *Fundamental relation for magnetic systems*

It remains for us to incorporate magnetic phenomena into the fundamental relation of thermodynamics. We begin with the differential form in the energy representation

$$dU = TdS - PdV + \mu dN + \Delta S_{mixing} + dW_{magn} \qquad (5.85)$$

We recall from electromagnetism that magnetizing a magnetic system reversibly by increasing the magnetic induction by $d\mathbf{B}$, requires work (Faraday's law)

$$dW_{magn} = - \int (\mathbf{H} \cdot d\mathbf{B}) dV \qquad (5.86)$$

integrated over all space. Written in this way, we can account for inhomogeneous, non-uniform fields and materials. Note that the differential is indeed extensive as $\mathbf{B}$ depends on the volume of the system. Likewise, $\mathbf{H}$ is the external control variable. Eliminating **B** in favor of the magnetization with (5.78) yields, for a uniform external field

$$d(U - U_0) = TdS - PdV + \mu dN + \mathbf{B}_0 \cdot d\mathbf{M} \tag{5.87}$$

Here

$$U_0 = \frac{1}{2} \int \mathbf{H} \cdot \mathbf{B}_0 dV \tag{5.88}$$

is the energy of the magnetic field in empty space (which is of no relevance when dealing with the thermodynamics of a system).

In closing we note that a similar modification of the fundamental relation can be done for electrically polarizable media in which case we get

$$d(U - U_0) = TdS - PdV + \mu dN + \mathbf{B}_0 \cdot d\mathbf{M} + \mathbf{D}_0 \cdot d\mathbf{P} \tag{5.89}$$

$$U_0 = \frac{1}{2} \int (\mathbf{H} \cdot \mathbf{B}_0 + \mathbf{E} \cdot \mathbf{D}_0) dV \tag{5.90}$$

where **P** is the electric dipole moment of the system.

5.8 Problems

Problem 5.1. *Calculate α, κ_T, c_V, and c_P for an ideal gas with $U = cNRT$ and $PV = NRT$.*

Problem 5.2. *(continuation of Problem 3.2) The entropy of N moles of a real (not ideal) gas has been determined to be*

$$S = Ns_0 + NR\ln[(v - b)(u + a/v)^c] \tag{5.91}$$

where $v = V/N$, $u = U/N$ and s_0 is a constant.
Calculate α, κ_T, c_V, and c_P.

Problem 5.3. *We know that thermal dissociation of hydrogen occurs around $T = 3500$ K under atmospheric pressures (see Figure 5.3), and that the molecular binding energy is roughly -4.5 eV. Despite this, the vast majority of hydrogen in the Universe is thought to be in the form of isolated hydrogen atoms, not molecules. Remember that the temperature of the Universe is a cool 3 K. How can this be understood?*

Problem 5.4. *(continuation of Problem 1.3) The Universe is considered to be an expanding blackbody cavity containing radiation that now is at a temperature of 2.7 K. Cosmological show that this expansion is isentropic.*

(a) *What was the temperature when the Universe had a diameter of only 1 cm, i.e. shortly after the Big Bang?*

(b) *What was the radiation pressure then and what is it now?*

(c) *What is the ratio of the kinetic energy of matter to the energy of radiation in intergalactic space?*

(d) *What is the ratio of the total matter energy (i.e. the sum of the kinetic energy plus the relativistic energy mc^2) to the energy of radiation?*

Problem 5.5.

(a) *How accurate is Stirling's approximation (5.75) for $N = 10, 100$?*

(b) *For which value of θ is the entropy (5.76) a minimum? For which value is it a maximum? Interpret this result in terms of the interpretation of the entropy as a measure of disorder.*

Problem 5.6. *Calculate and plot the isothermal compressibility of the Wormlike chain (5.39) as a function of L/L_c.*

Problem 5.7. *A piece of material known to superconduct (type I) at low T is placed in a beaker at room temperature. The beaker is then immersed in liquid nitrogen ($T < T_c$). A rare earth magnet is then lowered into the beaker from above. Because the superconductor expels magnetic fields, the magnet hovers above its surface at a height h. A second experiment is then conducted where an identical superconductor and identical magnet are placed in a beaker at room temperature. Since $T > T_c$, the sample is not in the superconducting state, and therefore a magnetic field can penetrate it without any energetic cost—the bar magnet rests directly on the sample. In the next step, the beaker is again immersed in liquid nitrogen. Once the sample becomes superconducting, it expels the magnet field, and the magnet is lifted off of the surface to a height h. Although the final states of these two experiments seem to be the same, in the second case the "sample" has performed work on the bar magnet, equal to mgh. Where does this extra energy come from?*

Chapter Summary

(1) Second derivatives are introduced: expansion coefficient, compressibility, specific heat, and more.

(2) Simple systems are discussed in detail:

    (a) Mixtures of ideal gases,

    (b) Chemical reactions,

    (c) Blackbody radiation,

    (d) Polymers,

    (e) Adsorbates, and

    (f) Magnetic systems.

Chapter 6

Thermodynamic Potentials

So far we have based thermodynamics on two fundamental relations, namely the entropy representation, $S = S(U, V, N)$, and the internal energy representation, $U = U(S, V, N)$. Although they are completely equivalent we have seen in a number of examples that one or the other may sometimes be more appropriate or convenient. It is not really surprising that we have two equivalent formulations of thermodynamics; other fields of physics do the same. Consider classical mechanics: we have (1) Newton's equations of motion, (2) Lagrange equations, (3) Hamilton's equations and (4) the Hamilton-Jacobi equation. Again, in which framework we work is a matter of convenience. Similarly in quantum mechanics we can work with Heisenberg's matrix mechanics or with Schrödinger's wave mechanics; again the choice is one of convenience. Geometry also offers a multitude of approaches: you can construct a circle by (1) minimizing the circumference of a fixed area, or (2) by maximizing the area for a fixed circumference. Both extremal principles yield circles. In this chapter we will successively eliminate the extensive variables in favor of their intensive counterparts. Eliminating the entropy from the internal energy will give us the Helmholtz free energy $F = F(T, V, N)$, further elimination of the volume will produce the Gibbs free energy $G = G(T, P, N)$, etc.

Let us then revisit the maximum entropy principle and deduce its implications for the internal energy. We start with an isolated composite system of fixed internal energy U, volume V, and mole number N. We separate off a subsystem by a rigid, adiabatic, and impermeable wall; the subsystem has variables U_1, V_1, and N_1. If we make the separating wall moveable, permeable, and diathermal, then these variables will adjust in such a way that the entropy is maximum, i.e. for $X = U_1$, V_1, and N_1, we have

$$\frac{\partial S}{\partial X}\Big|_U = 0$$

$$\frac{\partial^2 S}{\partial X^2}\Big|_U < 0 \tag{6.1}$$

What does that imply for U? We have

$$\frac{\partial U}{\partial X}\Big|_S = -\frac{\frac{\partial S}{\partial X}\big|_U}{\frac{\partial S}{\partial U}\big|_X} = -T\frac{\partial S}{\partial X}\Big|_U = 0 \tag{6.2}$$

Thus U is also an extremum over the constrained states but for fixed entropy. So take another derivative and you find

$$\frac{\partial}{\partial X}\frac{\partial U}{\partial X}\Big|_S = -T\frac{\partial^2 S}{\partial X^2}\Big|_U > 0 \tag{6.3}$$

Thus we have two extremum principles.

Entropy Maximum Principle: The equilibrium value of any unconstrained internal parameter is such that entropy is maximized for a given value of the total internal energy.

Energy Minimum Principle: The equilibrium value of any unconstrained internal parameter is such that the internal energy is minimized for a given value of the total entropy.

A graphical depiction of this dualism is shown in Figure 6.1.

Are there more convenient thermodynamic potentials? That is to say, are there potentials that keep other variables fixed. Consider adiabatic expansion for which the entropy is fixed. Obviously it is very convenient to use the internal energy, because for fixed entropy and fixed mole number it is a function of volume only; it is always easier to work with a function of one variable. What about isothermal expansion? What we would need is a thermodynamic potential that has temperature and volume as independent parameters and yet contains all the thermodynamic information about the system.

6.1 Introducing internal constraints via reservoirs

Up to this point we have discussed systems isolated from their environment so that we could characterize them by specifying the values for their extensive variables U, V, N_i, S. We have also looked at composite systems separated from one another by walls with different transport characteristics. More often than not, however, systems are not isolated from their environment but rather are intricately coupled to it. As an example, if we perform

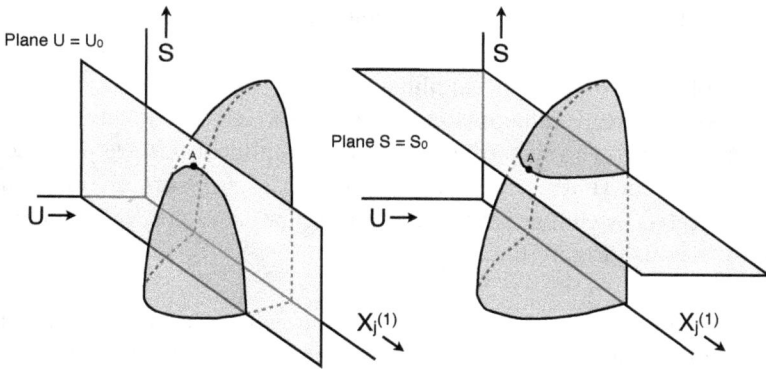

Fig. 6.1 Extremum Principle: The equilibrium state of a system is a point of maximal entropy for constant internal energy (left) OR a point of minimal internal energy for constant entropy (right). Image adapted from [Callen (1985)]

a chemical reaction in an open Petri dish, pressure is controlled by the environment. What happens in our experiment will not alter the pressure of the room. Indeed, there are many reactions we would not want to perform within a closed vessel for fear of an explosion. Likewise, we might want to cool a device to avoid overheating. In that situation the temperature is controlled by the environment. For any such situation where a system is coupled or in contact with its environment, it would be advantageous to re-formulate thermodynamics so that the coupling is taken care of.

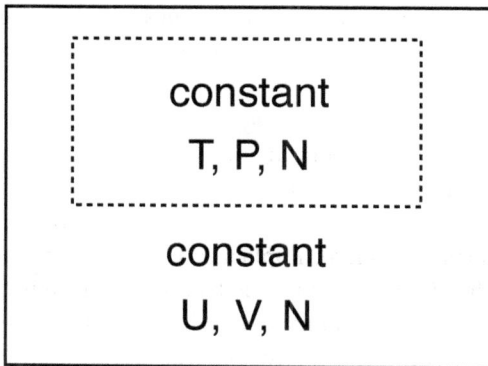

Fig. 6.2 A "system" in contact with a much larger system, the reservoir.

If the environment to which the system is intricately coupled is of a size similar to the system itself, then the only way to account for the coupling is by considering the two systems as a supersystem that is isolated from the rest of the world. No simplification is possible in this case, as we have already seen in numerous examples. However, if the environment is large compared to the system under study, simplifications are possible. By "large" we mean that certain properties such as pressure or temperature are not affected by changes in the system. A few examples will demonstrate the idea, see also Figure 6.2.

(a) A gas in a cylinder fitted with a piston: when the piston is fixed the volume of the system remains constant and we have to calculate the pressure as the derivative of the internal energy. However, if we make the piston moveable and put a weight of mass M on it, the pressure stays constant at $P = Mg/A$, where A is the area of the piston. Calculations would then be simplified if we had a thermodynamic function or **potential** in which the pressure is an argument, and the volume changes in the system would be calculated as a derivative of this pressure. We call an auxiliary system with constant pressure a **pressure reservoir**. The atmosphere above an open Petri dish is another pressure reservoir.

(b) A box which can be maintained at a constant temperature by temperature controlled heating is a **temperature reservoir**. We can put our system inside it to maintain it at that temperature. Entropy changes would then be calculated as a derivative of this new potential with respect to the temperature. The furnace of a power plant is a hot temperature reservoir whereas the cooling water from a river or lake constitutes a cold reservoir.

Coupling a system to a reservoir implicitly enforces a constraint on the previously isolated system and creates a super system consisting of the actual system and the reservoir. Because of this constraint, the principle of entropy maximization no longer applies to the system itself, but rather the super system. What is now up to us is to find a mathematical formulation that imposes such constraints automatically, namely those new thermodynamic potentials with internal constraints as natural variables. This leads to new extrema principles that do not involve the external reservoir or reservoirs. Starting from the internal energy, $U = U(S, V, N_1, N_2, ...)$, we eliminate successively the entropy in favor the temperature, the volume in favor of the pressure, and the mole numbers in favor of the chemical potentials.

6.2 Helmholtz free energy

We start from the differential form of the internal energy
$$dU = TdS - PdV + \mu dN \tag{6.4}$$
and write
$$TdS = d(TS) - SdT \tag{6.5}$$
to get
$$dU = d(TS) - SdT - PdV + \mu dN \tag{6.6}$$
or
$$d(U - TS) = -SdT - PdV + \mu dN \tag{6.7}$$

We call $F = U - TS$ the **Helmholtz free energy** or the Helmholtz potential; its differential form is
$$dF = -SdT - PdV + \mu dN \tag{6.8}$$
and thus it is a function of temperature, volume, and mole number
$$F = F(T, V, N) = U(S(T, V, N), V, N) - TS(T, V, N) \tag{6.9}$$

Wherever the entropy appears it must be replaced by its functional dependence on T, V, and N. Taking partial derivatives we can write down the differential form
$$dF = \frac{\partial F}{\partial T}|_{V,N}dT + \frac{\partial F}{\partial V}|_{T,N}dV + \frac{\partial F}{\partial N}|_{V,T}dN \tag{6.10}$$
and by comparison with (6.8) get the three equations of state in the Helmholtz representation
$$S = -\frac{\partial F}{\partial T}|_{V,N} = S(T, V, N)$$
$$P = -\frac{\partial F}{\partial V}|_{T,N} = P(T, V, N)$$
$$\mu = \frac{\partial F}{\partial N}|_{V,T} = \mu(T, V, N) \tag{6.11}$$

The transformation from $U(S, V, N)$ to $F(T, V, N) = U - TS$ is the first of a number of **Legendre** transformations in thermodynamics; it is also used in the Hamiltonian formulation of classical mechanics and in geometry to explore the duality between points and lines.

Remark 6.1. S obtained as a derivative of the Helmholtz free energy is a function of T, V, N and is thus NOT a fundamental relation. It is one of the three equations of state.

In a system with fixed mole number a quasi-static process at fixed temperature leads to a change

$$dF = -PdV = đW \tag{6.12}$$

that is equal to the work done on the system. Turning the argument around, $-dF$ is the energy available at fixed temperature to do work on the outside, hence the term free energy.

Recall that to fix the temperature in a system, we must bring it into contact with another system that is so large that the addition or extraction of a small amount of energy will not change its temperature T_r; such a system is called a heat or thermal reservoir. The combined system (the system plus the heat reservoir) has an internal energy $U + U_r$, where the energy of the reservoir is simply $U_r = T_r S_r$ if we ensure that it is surrounded by fixed and impermeable walls. In equilibrium the internal energy and the entropy of the combined system are extrema

$$d(U + T_r S_r) = 0$$
$$d(S + S_r) = 0 \tag{6.13}$$

Using the second equation and remembering that in equilibrium any systems in thermal contact are at equal temperature, $T = T_r$, we get

$$d(U - TS) = dF = 0 \tag{6.14}$$

Likewise we can show that the second derivative of F is positive so that the Helmholtz free energy is a minimum over the manifold of states for which T=T_r.

Example 6.1. A cylinder contains an internal piston, on each side of which is one mole of a monatomic ideal gas. The walls of the cylinder are diathermal. The cylinder is immersed in a large tub of water plus some ice serving as the heat reservoir at a temperature $T = 0°$ C. The initial volumes of the two gaseous subsystems are 10 liters and 1 liter. The piston is now moved quasi-statically, so that the final volumes are 6 liters and 5 liters, respectively. How much work is delivered?

From (5.15) we get the Helmholtz free energy as

$$F = NRT \left\{ \frac{F_0}{N_0 RT_0} - \ln \left[(\frac{T}{T_0})^{3/2} \frac{V}{V_0} \frac{N_0}{N} \right] \right\} \tag{6.15}$$

or, for fixed temperature and mole numbers, $F = constant - NRT \ln V$, and

$$\Delta F = -NRT \ln(\frac{V_1^{(f)} V_2^{(f)}}{V_1^{(i)} V_2^{(i)}}) = -NRT \ln 3 = -2.5 kJ \tag{6.16}$$

This energy is extracted from the heat reservoir and is made available to do work on the outside. In an isothermal process the internal energy of an ideal gas does not change ($U = cNRT$) but its entropy has changed by $\Delta S = -\Delta F/T$.

6.3 Enthalpy

The procedure by which we constructed the Helmholtz free energy for isothermal processes can also be applied to any of the other arguments of the internal energy. Next is the elimination of the volume in favor of the pressure. We use the identity $PdV = d(PV) - VdP$ and get

$$dH = d(U + PV) = TdS + VdP + \mu dN \tag{6.17}$$

and get for the **enthalpy** as a function of its natural variables S, P, and N

$$H = H(S, P, N) = U(S, V(S, P, N)) + PV(S, V(S, P, N)) \tag{6.18}$$

Wherever the volume appears it must be replaced by its functional dependence on S, P, and N. Taking its partial derivatives we can write down the differential form

$$dH = \frac{\partial H}{\partial S}|_{P,N}dS + \frac{\partial H}{\partial P}|_{S,N}dP + \frac{\partial H}{\partial N}|_{S,P}dN \tag{6.19}$$

and by comparison with (6.17) get the three equations of state in the enthalpy representation

$$S = \frac{\partial F}{\partial T}|_{P,N} = S(T, P, N)$$

$$V = \frac{\partial H}{\partial V}|_{S,N} = V(S, P, N)$$

$$\mu = \frac{\partial H}{\partial N}|_{S,P} = \mu(S, P, N) \tag{6.20}$$

One application to which the enthalpy is ideally suited is the description of chemical reactions in open vessels. While pressure is constant (and is equal to that of the lab), temperature changes because during reactions heat is either produced or consumed. For any system in contact with a pressure reservoir ($P = $ constant) and no chemical reactions (constant mole numbers) heat transfer results in an increase of enthalpy $dH = đQ$. The enthalpy is perfect for the description of the Joule-Thomson "throttling" process. This process can be used to reach low temperatures (and thereby liquify gases).

6.3.1   *Joule-Thomson "throttling" process*

The experiment is very simple. A gas at high pressure seeps through a porous plug into a region of low pressure. In Joule's original experiment, the porous plug was just a wad of cotton stuffed into a pipe. High and low pressures are maintained by moveable pistons (see Figure 6.3).

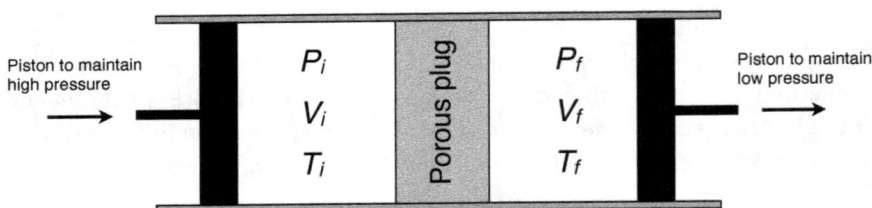

Piston to maintain high pressure

P_i

V_i

T_i

Porous plug

P_f

V_f

T_f

Piston to maintain low pressure

Fig. 6.3   Schematic of the Joule-Thomson throttling experiment.

During this experiment, the temperature of the gas is not constant. The fact that the gas temperature changes is not surprising in and of itself; recall that a gas cools in adiabatic expansion (and heats during adiabatic compression). Unlike adiabatic processes though, during throttling a gas is either heated or cooled depending on its initial temperature.

Let us assume that initially we have a gas on the left side of the throttle at pressure P_i, temperature T_i, and molar volume v_i. We then push the whole amount of gas through the throttle to the low pressure side with P_f, T_f, and v_f. The work required to reduce the initial volume to zero at constant pressure is

$$-\int_{v_i}^{0} P_i dv = P_i v_i \qquad (6.21)$$

and similarly for the expansion on the right hand side. Thus the final internal energy per mole is

$$u_f = u_i + P_i v_i - P_f v_f \qquad (6.22)$$

or

$$u_i + P_i v_i = u_f + P_f v_f \qquad (6.23)$$

$$h_f = h_i \qquad (6.24)$$

In an actual machine, throttling is done at constant enthalpy, but at a rate so fast that the intermediate states are out of equilibrium. That seems to indicate that we can not use thermodynamics to describe this process. However, we can reach the same final state of equal enthalpy by pushing the gas slowly through the plug so that we can perform the throttling process quasi-statically at constant enthalpy using thermodynamics for its description. To learn more about the process we want to calculate how much the temperature changes for a small pressure change dP, i.e.

$$dT = \frac{\partial T}{\partial P}|_{H,N}dP \tag{6.25}$$

$$= -\frac{(\partial H/\partial P)_T}{(\partial H/\partial T)_P}dP$$

Using the differential for the enthalpy at constant mole number $dH = TdS + VdP$, the numerator and denominator become

$$\frac{\partial H}{\partial P}|_T = T\frac{\partial S}{\partial P}|_T + V \tag{6.26}$$

$$\frac{\partial H}{\partial T}|_P = T\frac{\partial S}{\partial T}|_P = Nc_P \tag{6.27}$$

We quote here without proof (to be given in our presentation of Maxwell's relations in the next chapter) the identity $(\partial S/\partial P)_T = (\partial V/\partial T)_P = V\alpha$. Thus the equation ruling the Joule-Thomson effect becomes

$$dT = \frac{v}{c_P}(T\alpha - 1)dP \tag{6.28}$$

Clearly, for $T\alpha > 1$ a small decrease in pressure will cool the gas, whereas for $T\alpha < 1$ it heats up. The inversion temperature is determined by

$$\alpha T_{inv} = 1 \tag{6.29}$$

and is in general pressure dependent. For an ideal gas the expansion coefficient is $\alpha = 1/T$ so there is no temperature change in an isenthalpic expansion. This is not surprising as the ideal gas model neglects interactions between the gas particles which, as a consequence, can never liquefy. As a gas expands, the average distance between molecules grows. In a real gas, because of intermolecular attractive forces between molecules, expansion causes an increase in the potential energy of the gas. If no external work is extracted in the process and no heat is transferred, the total energy of the gas remains the same because of the conservation of energy. The increase in potential energy thus implies a decrease in kinetic energy and therefore in temperature.

A second mechanism has the opposite effect. During gas molecule collisions, kinetic energy is temporarily converted into potential energy. As the average intermolecular distance increases, there is a drop in the number of collisions per time unit, which causes a decrease in average potential energy. Again, if total energy is conserved, this leads to an increase in kinetic energy (temperature). Below the Joule–Thomson inversion temperature, the former effect (work done internally against intermolecular attractive forces) dominates, and free expansion causes a decrease in temperature. Above the inversion temperature, gas molecules move faster and so collide more often, and the latter effect (reduced collisions causing a decrease in the average potential energy) dominates: Joule–Thomson expansion causes a temperature increase.

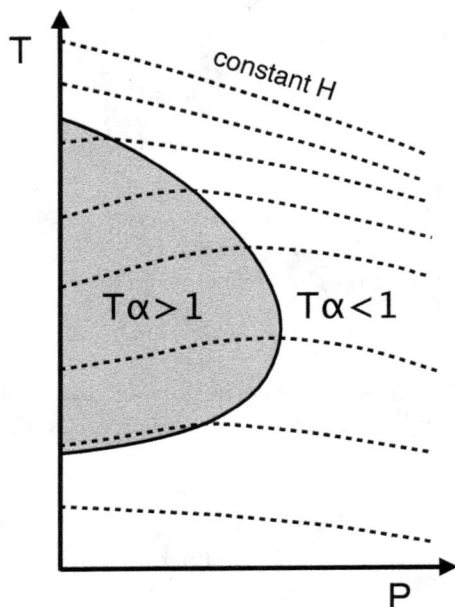

Fig. 6.4   Isenthalps for the Joule-Kelvin throttling experiment. For initial states in the region $T\alpha > 1$ cooling occurs.

In Figure 6.4 we show the isenthalps for a gas in the temperature-pressure diagram; they are concave and their maxima define the inversion temperature as a function of pressure. To the left of the inversion curve the Joule-Thomson throttling process cools a gas, to the right it heats up. The maximum inversion temperatures of some common gases together with their boiling points at atmospheric pressure are listed in the table.

Gas	Xe	O_2	CO	N_2	Ne	H_2	^{4}He
maximum T_{inv} [K]	1486	764	644	621	228	204	43
boiling $T_{boiling}$ [K]	161.2	90.18	81	77.35	24.5	20.27	4.21

6.3.2 Gas liquefaction

A schematic of the Linde gas liquefaction machine is shown in Figure 6.5. A gas below its inversion temperature is compressed to high pressure and sent down a tube to a "plug" which is nothing but a small orifice at the end of the tube (if the tube is made from copper, just squeezing the end with a pair of pliers will do). The gas will expand isenthalpically and cool down. This cooled gas is sent back to the compressor, cooling the already compressed gas on its way to the orifice. After many cycles the gas will be so cold that it liquefies; for nitrogen this happens at 77 K or −196° C.

Remark 6.2 (history). *Carl Paul Gottfried von Linde (1842 - 1934) was a German engineer who developed refrigeration and gas separation technologies. Born in Bavaria, he started his studies in engineering in 1861 at the Swiss Federal Institute of Technology in Zürich, Switzerland, where his teachers included Rudolf Clausius (the thermodynamicist who introduced entropy). After several jobs in industry he became a professor of engineering at the Technical University of Munich. This is where he did his early work on refrigeration using the Joule-Thomson effect; Rudolf Diesel, the inventor of the Diesel engine, was one of his prominent students. Linde's first refrigeration plants were commercially successful, and he founded the Gesellschaft für Linde's Eismaschinen Aktiengesellschaft ("Linde's Ice Machine Company"), now Linde AG, in Wiesbaden, Germany. The efficient new refrigeration technology offered big benefits to the breweries, and by 1890 Linde had sold 747 machines. Other uses for the new technology were found in slaughterhouses and cold storage facilities all over Europe. In succession he liquified air, oxygen, nitrogen, and acetylene (for torch welding). In addition to Linde's technical and engineering abilities, he was a successful entrepreneur. He also formed the Linde Air Products Company in the USA in 1907, a company that passed through US Government control to Union Carbide in the 1940s. It would become Praxair. Further applications for the new refrigeration technology were, among others, ice rinks, chilling and freezing units for ships and railway cars, units for air humidification and refrigeration in residential areas as well as industrial units for chilling milk in dairies or for refrigeration in sugar and chocolate factories.*

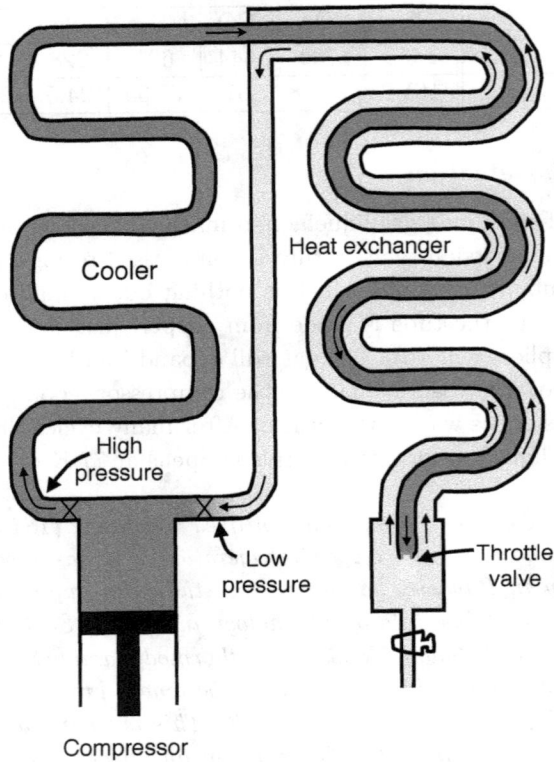

Fig. 6.5   Apparatus to liquify gases.

In 1932, the Linde company supplied the first industrial helium lique-faction plant in the world to the University of Kharkov in the Ukraine. He-lium liquefies at a temperature of 272.2 degrees below zero Celsius, in other words close to the absolute zero point (273.13 degrees below zero Celsius). In collaboration with Prof. Walther Meissner, Linde developed a plant that combined the Linde air liquefaction process with a pre-refrigeration of the helium using liquid hydrogen. Today Linde has virtually no competition in the area of achieving the lowest temperatures (cryogenics) and ranks world-wide among the technology leaders. With more than 1500 process en-gineering patents and 3500 completed plant projects Linde ranks among the leading international plant contractors.

Moral of the story: You can make money with thermodynamics.

Fig. 6.6   Linde's first refrigerator, the "ice machine".

6.4   Gibbs free energy

We return now to the construction of new thermodynamic potentials, combining the Legendre transformation from the internal energy (into the Helmholtz free energy) with that of the enthalpy. This new function is called the Gibbs free energy and is a function of two intensive variables T and P, and N as the remaining extensive variable

$$dG = -SdT + VdP + \mu dN$$

$$S(T,P,N) = -\frac{\partial G}{\partial T}\Big|_{P,N}$$

$$V(T,P,N) = -\frac{\partial G}{\partial P}\Big|_{T,N} \tag{6.30}$$

Note also that

$$G = U - TS + PV = N\mu(T,P) \tag{6.31}$$

so that the molar Gibbs free energy $g = G/N = \mu(T, P)$. It is the perfect potential to deal with chemical reactions. Because we have already looked into chemical reactions in the previous chapter, we give here as a different example of the usefulness of the Gibbs free energy, a description of the vapor pressure of a small liquid droplet.

6.4.1 *Vapor pressure of small droplets*

Let us study an isolated liquid droplet of radius r in equilibrium. We cut away the lower half of the droplet and ask which forces must act on that half-droplet so that it remains in stable equilibrium, see Figure 6.7.

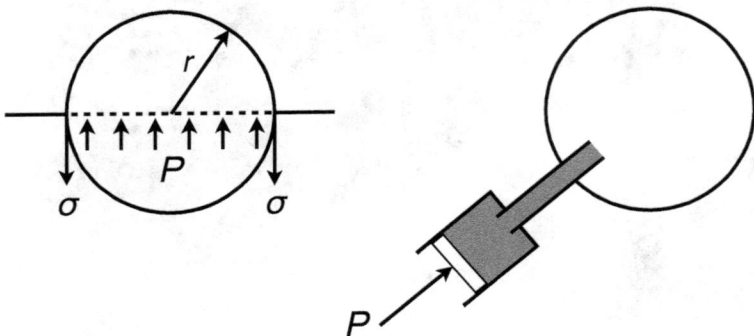

Fig. 6.7 Vapor pressure of a droplet. Left: force balance on half a droplet; right: injecting extra liquid into a droplet.

If originally a pressure P acted on the interior of the whole droplet we must apply a force $\pi r^2 P$ upward, as did the lower half of the droplet before surgery. Additionally, by cutting the lower half away we have created a rim of length $2\pi r$. On a length segment dl a force γdl must act downward, again to counteract the removal of the lower half; here γ is the surface tension. The total force acting on the whole cut is therefore $2\pi r\gamma - \pi r^2 P$ and this force must be zero for the upper half alone to be stable. Therefore we must have in the interior of the droplet a pressure

$$P = \frac{2\gamma}{r} \tag{6.32}$$

which depends on the size of the droplet.

Let us take a moment to look at this result. Starting with a droplet of volume V, we increase it by dV by injecting some more liquid into the droplet. Doing this isothermally (the droplet is in thermal contact with the vapor around it!) we do work PdV, increasing its Helmholtz free energy by that amount. But this increase only originates from the increase in the surface area of the droplet, $dF = \gamma dA$. Thus we have $PdV = \gamma dA$. With $A = \pi r^2$ and $V = r^3 4\pi/3$, we get again (6.32).

A droplet is only stable, i.e. does not shrink or grow in size, if it is in contact with its own vapor pressure. Its dependence on the radius of the droplet is what we want to calculate next. We first do this in the framework of the Gibbs free energy which we write as $G = F + PV$. Note that in a liquid atoms or molecules are closely packed so that the volume of a mole of a liquid is roughly $v_L = N_A v_a$, where N_A is Avogadro's number and v_a is the volume of one atom or molecule. We then get for a droplet consisting of N moles

$$G = F + 4\pi r^2 \sigma + N v_L P_r \tag{6.33}$$

The second term accounts for the surface free energy, and in the last term p_r is the yet unknown vapor pressure of the droplet under which it is stable. From this we get the chemical potential

$$\mu_{droplet} = \frac{\partial G}{\partial N} = F/N + v_L P_r + 4\pi\sigma \frac{\partial r^2}{\partial N} \tag{6.34}$$

For the last term we note that $N v_L = 4\pi r^3/3$ or $r^2 = (3N v_L/4\pi)^{2/3}$ and get

$$\mu_{droplet} = F/N + v_L P_r + \frac{2 v_L \sigma}{r} \tag{6.35}$$

On the other hand, the vapor around the droplet is at low density and can be approximated by an ideal gas for which

$$\mu_{vapor} = RT \ln P_r + B(T) \tag{6.36}$$

In equilibrium the chemical potentials are equal

$$\mu_{droplet} = \mu_{vapor}$$

$$F/N + v_L P_r + \frac{2 v_L \sigma}{r} = RT \ln P_r + B(T) \tag{6.37}$$

For a plane surface ($r \to \infty$) this condition reads

$$F/N + v_L P_\infty = RT \ln P_\infty + B(T) \tag{6.38}$$

where P_∞ is the saturation vapor pressure over a plane liquid surface. Subtracting these two equations we get

$$\frac{RT}{v_L} \ln \frac{P_r}{P_\infty} = \frac{2\sigma}{r} + (P_r - P_\infty) \tag{6.39}$$

This can be further simplified on numerical grounds. Write

$$P_\infty \left[\frac{RT}{P_\infty v_L} \ln \frac{P_r}{P_\infty} + 1 - \frac{P_r}{P_\infty} \right] = \frac{2\sigma}{r} \qquad (6.40)$$

Now RT/P_∞ is the volume of one mole of vapor which is typically 3 to 4 orders of magnitude larger than that of one mole of liquid; thus the log-term on the left dominates and we get

$$\ln \frac{P_r}{P_\infty} = \frac{M}{\rho RT} \frac{2\sigma}{r} \qquad (6.41)$$

where M is the molar weight and ρ is the mass density of the liquid. If $P_r < P_\infty$ the left side is negative and no droplets can exist. If $P_r > P_\infty$ the vapor is supersaturated and droplets start to form; fog and clouds are typical examples. Because droplets are usually small and hard to catch and to measure, thermodynamics is very helpful—it allows us to calculate the droplet size from measurements of the supersaturated vapor pressure!

6.4.2 *Osmosis*

Osmosis is the diffusion of a solvent (typically water) through a semi-permeable membrane without input of energy. The membrane is permeable to the solvent but not the solute (e.g. sugar or salt). Osmosis releases energy, and can be made to do work, as for example when a growing tree root splits a stone.

Osmosis is extremely important in biological systems. Many membranes are semipermeable. In general, these membranes are impermeable to organic solutes with large molecules, such as polysaccharides, while permeable to water and small, uncharged solutes. Permeability may depend on solubility properties, charge, or chemistry, in addition to the solute size. Water molecules travel through the plasma cell wall, tonoplast (vacuole) or protoplast in two ways, either by diffusing across the phospholipid bilayer directly, or via aquaporins (small transmembrane proteins similar to those that create ion channels). Osmosis provides the primary means by which water is transported into and out of cells. Furthermore, the turgor pressure between the cell interior and its relatively hypotonic environment is largely maintained by osmosis across the cell membrane. In many plants, the ability to stand upright is due to osmotic processes. Osmosis is also responsible for the regulation of the aperture of stomata within their outer skin.

Reverse osmosis is a separation process that uses pressure to force a solvent through a membrane that retains the solute on one side and allows the pure solvent to pass to the other side. It is the basis of filtering, a process commonly used to purify water. The water to be purified is placed in a chamber and put under an amount of pressure greater than the osmotic pressure exerted by the water and the solutes dissolved in it. Part of the chamber opens to a differentially permeable membrane that lets water molecules through, but not the solute particles. The osmotic pressure of ocean water is about 27 atm. Reverse osmosis desalinators use pressures around 70 atm to produce fresh water from ocean salt water.

To understand osmosis, we require a thermodynamic potential for a mixture of solvent and solute. Because we want to determine the osmotic pressure difference at constant temperature, the Gibbs free energy seems appropriate. Considering only dilute solutions we can write for N_1 moles of solvent and N_2 moles of solute

$$G(T, P, N_1, N_2) = N_1 \mu_1^0(T, P) + N_2 \alpha(T, P)$$
$$+ N_1 RT \ln\left(\frac{N_1}{N_1 + N_2}\right)$$
$$+ N_2 RT \ln\left(\frac{N_2}{N_1 + N_2}\right) \quad (6.42)$$

Here $\mu_1^0(T, P)$ is the chemical potential of the pure solvent and $\alpha(T, P)$ is its modification due to the solute. We have added a term TS_{mixing} to account for mixing, see Section 5.2. For a dilute mixture, the mole fraction of solute is small, $c = N_2/(N_1 + N_2) \simeq N_2/N_1 \ll 1$, simplifying the Gibbs free energy accordingly. Derivatives with respect to the mole numbers give the respective chemical potentials

$$\mu_1(T, P, c) = \frac{\partial G}{\partial N_1}|_{T,P} = \mu_1^0(T, P) - cRT$$
$$\mu_2(T, P, c) = \frac{\partial G}{\partial N_2}|_{T,P} = \alpha(T, P) + RT \ln c \quad (6.43)$$

In our first example we look at a U-tube which is separated into two halves by a semipermeable wall, permeable to the solvent only, see Figure 6.8.

Initially, we fill both sides of the tube to the same height, where the liquid on the right is pure water and the liquid on the left is a mixture of sugar and water. Because the solute cannot permeate into the right half, we need only consider equilibrium of the solvent on both sides

$$\mu_1^r(T, P^r, 0) = \mu_1^l(T, P^l, c) \quad (6.44)$$

or

$$\mu_1^r(T, P^r, 0) = \mu_1^l(T, P^l, 0) - cRT$$

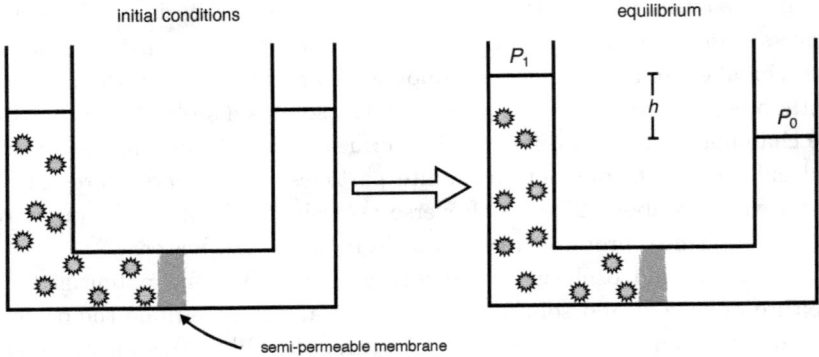

Fig. 6.8   The presence of a semi-permeable membrane can result in a chemical concentration gradient that would not be possible otherwise. In this case, the presence of impurities on the left hand side of the cell reduces the chemical potential of water. The result is an osmotic pressure. This pressure is sufficiently large that when studying living cells under a microscope, a saline solution must be used. If it is not, the cells will absorb water and eventually burst.

For a dilute solution the pressure change is not large, so we can use a Taylor expansion on the right hand side

$$\mu_1(T, P^l, 0) = \mu_1(T, P^r, 0) + \frac{\partial \mu_1}{\partial P}(P^l - P^r)$$

$$= \mu_1(T, P^r, 0) + \frac{V}{N_1}(P^l - P^r) \tag{6.45}$$

Inserted into the previous equation this gives, after multiplication with N_1

$$V\Delta P = N_2 RT \tag{6.46}$$

This is the van't Hoff equation for the osmotic pressure in dilute solutions. Note that the pressure in the solution on the left hand side is larger than in the pure solvent. This pressure increase is maintained by raising the liquid level by an amount h given

$$h\rho g + P_1 - P_0 = \Delta P \tag{6.47}$$

where P_0 and P_1 are the vapor pressures over the pure solvent on the right and the solution on the left. They are connected by the barometric height formula

$$\frac{P_1}{P_0} = \exp\left(-\frac{Mgh}{RT}\right) \tag{6.48}$$

where M is the molar mass of the solvent/solution. Thus the osmotic pressure can be simply calculated from the measured value of h.

A variant of this experiment is the following: instead of a U-tube, consider the case where we we have two open containers: one containing pure solvent and the other containing the mixture (solution). If these two vessels are placed in the vicinity of one another in a closed room, what will happen? As with the previous case, it is favorable to dilute the solution by moving solvent. Since there is no connecting membrane, there is only one way to achieve this: evaporation. Liquid will evaporate out of the solvent container and will condense into the solute one. While the method of transport in this experiment is different, the thermodynamic driving force is essentially the same.

6.5 Problems

Problem 6.1. *Find the fundamental equation for an ideal gas in the Helmholtz, enthalpy, and Gibbs representation. In each case calculate the three equations of state by differentiation.*

Problem 6.2. *The enthalpy of a particular system is given by*

$$H(S, P, N) = AS^2 n^{-1} \ln \left(\frac{P}{P_0} \right)$$

Calculate the specific heat at constant volume

$$c_V = \frac{T}{N} \frac{\partial S}{\partial T} \bigg|_{V,N}$$

as a function of T and P.

Problem 6.3. *For a particular system it is found that ($u = U/N$, $v = V/N$)*

$$u = \frac{3}{2} Pv$$
$$P = AvT^4$$

Find the molar Gibbs free energy and the molar Helmholtz free energy.

Problem 6.4. *A gas has the following equations of state*

$$P = U/V$$
$$T = 3B(U^2/NV)^{1/3}$$

(a) *Find the molar entropy and show that it obeys Nernst's theorem ($S \to 0$ when $T \to 0$).*

(b) *The gas, initially at T_i and P_i is passed through a porous plug in a Joule-Thomson process. Find the final temperature T_f when the final pressure is P_f.*

Problem 6.5. *Using the concept of energy minimization, derive Archimedes Principle, which states that the buoyant force on a submerged object is equal to the weight of the fluid that is displaced by the object.*

Problem 6.6. *Consider a 1-dimensional chain of charged particles, interacting through a repulsive Coulomb potential. Calculate the total energy of this chain, and derive the $f(L)$ equation of state for $T = 0$ K, where L is the length of the chain and f is the applied force. Use these results to write down an expression for the enthalpy, H.*

Chapter Summary

We summarize the results of this chapter in the form of a table.

Legendre transform	Potential	variables	differential
	Entropy S	U, V, N_k	$dS = \frac{1}{T}dU + \frac{P}{T}dV + \sum_k \frac{\mu_k}{T}dN_k$
	Energy U	S, V, N_k	$dU = TdS - PdV + \sum_k \mu_k dN_k$
$U_T = U - TS$	Helmholtz F	T, V, N_k	$dF = -SdT - PdV + \sum_k \mu_k dN_k$
$U_P = U + PV$	Enthalpy H	S, P, N_k	$dH = TdS + VdP + \sum_k \mu_k dN_k$
$U_{T,P} = U_T + PV$	Gibbs G	T, P, N_k	$dG = -SdT + VdP + \sum_k \mu_k dN_k$
$U_{T,\mu} = U_T - \mu N$	Grand L	T, V, μ_k	$dL = -SdT - PdV - \sum_k N_k d\mu_k$
$U_{T,P,\mu} = U_{T,P} - \mu N$	Total K	T, P, μ_k	$dK = 0$

The grand potential L is well suited to study systems in a fixed volume and at a fixed temperature, such as a gas in equilibrium with its adsorbate on a solid surface, or a two phase region with some of the fluid in the gas phase and some in the liquid. In both cases what is not known are the mole numbers in the two phases, which you get by equating chemical potentials.

The total potential $K = U - TS + PV - \Sigma_k \mu_k N_k \equiv 0$ is still useful as it is an implicit constraint on all thermodynamic variables and potentials.

Chapter 7

Maxwell Relations

In the previous chapter we have introduced different thermodynamic potentials that are equivalent in their information content, yet prove useful in different applications. First derivatives of these potentials with respect to their natural variables give the equations of state for the conjugate variables. In this chapter we want to develop a strategy to efficiently deal with second derivatives. When we discussed the Joule-Thomson "throttling" process in Section 6.3.1 we came upon an expression (6.25)

$$dT = \frac{\partial T}{\partial P}\big|_{H,N} dP \tag{7.1}$$

and stated that (6.28)

$$\frac{\partial T}{\partial P}\big|_{H,N} = \frac{v}{c_P}(T\alpha - 1) \tag{7.2}$$

When dealing with such second derivatives a systematic approach to at least part of the task is based on Maxwell Relations.

7.1 Maxwell relations

The origin of Maxwell relations is very simple: consider a differentiable function of at least two variables, $f(x, y, z, ...)$. Its mixed second derivatives are symmetric

$$\frac{\partial^2 f}{\partial x \partial y} = \frac{\partial^2 f}{\partial y \partial x} \tag{7.3}$$

For clarity we recall what these second derivatives mean

$$\frac{\partial^2 f}{\partial x \partial y} = \frac{\partial}{\partial x}\left[\frac{\partial f}{\partial y}\big|_{x,z,...}\right]_{y,z,...} \tag{7.4}$$

For a function of n variables there are $n(n-1)/2$ pairs of such mixed second derivatives and each yields a Maxwell relation. And, because we have in the energy representation seven distinct thermodynamic potentials for a simple system with $n = 3$ (of which five are dominant in applications, U, F, H, G, and L) we get 21 Maxwell relations. Let us go through an example, starting with $U = U(S, V, N)$

$$\frac{\partial^2 U}{\partial S \partial V} = \frac{\partial}{\partial S} \frac{\partial U}{\partial V}\Big|_{S,N} = -\frac{\partial P}{\partial S}\Big|_{V,N} \tag{7.5}$$

and, starting from the reverse order of differentiation

$$\frac{\partial^2 U}{\partial V \partial S} = \frac{\partial}{\partial V} \frac{\partial U}{\partial S}\Big|_{V,N} = \frac{\partial T}{\partial V}\Big|_{S,N} \tag{7.6}$$

Thus equating the two we get our first Maxwell relation

$$\frac{\partial T}{\partial V}\Big|_{S,N} = -\frac{\partial P}{\partial S}\Big|_{V,N} \tag{7.7}$$

The procedure is now clear and we can summarize the results in tabular form.

$U(S, V, N)$	S, V	$\frac{\partial T}{\partial V}\Big|_{S,N} = -\frac{\partial P}{\partial S}\Big|_{V,N}$
$dU = TdS - PdV + \mu dN$	S, N	$\frac{\partial T}{\partial N}\Big|_{S,N} = \frac{\partial \mu}{\partial S}\Big|_{V,N}$
	V, N	$-\frac{\partial P}{\partial N}\Big|_{S,V} = \frac{\partial \mu}{\partial V}\Big|_{S,N}$

$$F(T,V,N) = U - TS \qquad T,V \qquad \frac{\partial S}{\partial V}\big|_{T,N} = \frac{\partial P}{\partial T}\big|_{V,N}$$

$$dF = -SdT - PdV + \mu dN \qquad T,N \qquad -\frac{\partial S}{\partial N}\big|_{T,V} = \frac{\partial \mu}{\partial T}\big|_{V,N}$$

$$V,N \qquad -\frac{\partial P}{\partial N}\big|_{T,V} = \frac{\partial \mu}{\partial V}\big|_{T,N}$$

$$H(S,P,N) = U + PV \qquad S,P \qquad \frac{\partial T}{\partial P}\big|_{S,N} = \frac{\partial V}{\partial S}\big|_{P,N}$$

$$dH = TdS + VdP + \mu dN \qquad S,N \qquad \frac{\partial T}{\partial N}\big|_{S,P} = \frac{\partial \mu}{\partial S}\big|_{P,N}$$

$$P,N \qquad \frac{\partial V}{\partial N}\big|_{S,P} = \frac{\partial \mu}{\partial P}\big|_{S,N}$$

$$U_\mu(S,V,\mu) = U - N\mu \qquad S,V \qquad \frac{\partial T}{\partial V}\big|_{S,\mu} = -\frac{\partial P}{\partial S}\big|_{V,\mu}$$

$$dU_\mu = TdS + VdP - Nd\mu \qquad S,\mu \qquad \frac{\partial T}{\partial \mu}\big|_{S,V} = -\frac{\partial N}{\partial S}\big|_{V,\mu}$$

$$V,\mu \qquad \frac{\partial P}{\partial \mu}\big|_{S,V} = \frac{\partial N}{\partial V}\big|_{S,\mu}$$

$$G(T,P,N) = U - TS + PV \qquad T,P \qquad \frac{\partial S}{\partial P}\big|_{T,N} = \frac{\partial V}{\partial T}\big|_{P,N}$$

$$dG = -SdT + VdP + \mu dN \qquad T,N \qquad -\frac{\partial S}{\partial N}\big|_{T,P} = \frac{\partial \mu}{\partial T}\big|_{P,N}$$

$$P,N \qquad \frac{\partial V}{\partial N}\big|_{T,P} = \frac{\partial \mu}{\partial P}\big|_{T,N}$$

$$U_{T,\mu}(T,V,\mu) = U - TS - \mu N \qquad T,V \qquad \frac{\partial S}{\partial V}\big|_{T,\mu} = \frac{\partial P}{\partial T}\big|_{V,\mu}$$

$$dU_{T,\mu} = -SdT - PdV - Nd\mu \qquad T,\mu \qquad \frac{\partial S}{\partial \mu}\big|_{T,V} = \frac{\partial N}{\partial T}\big|_{V,\mu}$$

$$V,\mu \qquad \frac{\partial P}{\partial \mu}\big|_{T,V} = \frac{\partial N}{\partial V}\big|_{T,\mu}$$

$$U_{P,\mu}(S,P,\mu) = U + PV - \mu N \qquad S,P \qquad \frac{\partial T}{\partial P}\big|_{S,\mu} = \frac{\partial V}{\partial S}\big|_{P,\mu}$$

$$dU_{P,\mu} = TdS + VdP - Nd\mu \qquad S,\mu \qquad \frac{\partial T}{\partial \mu}\big|_{S,P} = -\frac{\partial N}{\partial S}\big|_{P,\mu}$$

$$P,\mu \qquad \frac{\partial V}{\partial \mu}\big|_{S,P} = -\frac{\partial N}{\partial P}\big|_{S,\mu}$$

Facing a second derivative, how can you know that there is a Maxwell relation to relate it to another second derivative? The straightforward procedure would be to look it up in the table (provided you have the table with you). A second approach is a set of mnemonic diagrams, see for example Callen's book [Callen (1985)]. A third way is to follow the simple mathematics that leads to them. Here is a four-step approach illustrated for the example

$$\frac{\partial \mu}{\partial V}\Big|_{T,N} \tag{7.8}$$

(1) Look at the variables at the "bottom", i.e. the variables held fixed and the one with respect to which you take the derivative; in this case V, T, and, N.
(2) Identify the thermodynamic potential for which these variables are natural, here $F(T, V, N)$.
(3) Look at the variable on the "top", here it is μ, and ask: is it a conjugate variable for the thermodynamic potential. Yes, it is

$$\mu = \frac{\partial F}{\partial N}\Big|_{T,V} \tag{7.9}$$

(4) Thus

$$\frac{\partial \mu}{\partial V}\Big|_{T,N} = \frac{\partial^2 F}{\partial N \partial V} = \frac{\partial^2 F}{\partial V \partial N} = -\frac{\partial P}{\partial N}\Big|_{T,V} \tag{7.10}$$

This approach works for many derivatives, but there are important derivatives such as (6.25) for which no Maxwell relation exists and other procedures must be applied.

7.2   Reduction of derivatives

In Section 1.2, and again in Section 5.1, we introduced two second derivatives of physical significance, namely the expansion coefficient α, the isothermal compressibility κ_T, and the specific heat c_V at constant volume (and a few more). It will now be our task to develop a procedure to reduce all other derivatives to just three, namely α, κ_T, and the specific heat at constant pressure c_P. Before we do this, we re-write their definitions as

$$\alpha = \frac{1}{V}\frac{\partial V}{\partial T}\Big|_P = \frac{1}{V}\frac{\partial^2 G}{\partial T \partial P}$$

$$\kappa_T = -\frac{1}{V}\frac{\partial V}{\partial P}\Big|_T = -\frac{1}{V}\frac{\partial^2 G}{\partial P^2}\Big|_T$$

$$c_P = \frac{T}{N}\frac{\partial S}{\partial T}\Big|_P = -\frac{T}{N}\frac{\partial^2 G}{\partial T^2}\Big|_P \tag{7.11}$$

and note that these three quantities are all second derivatives of the Gibbs free energy, and the only ones for constant mole number. This "symmetry", together with their obvious physical significance has lead to the universal acceptance of these quantities as the basic ones, although any other choice would be just as good.

There is a recipe of five easy-to-follow rules that reduces any other derivative to an expression involving these three. Following this recipe avoids going in circles. What we need, in addition to the Maxwell relations and the differential forms for the thermodynamic potentials, is some basic results from the calculus of partial derivatives, in particular the following three rules (see also Appendix A):

$$\frac{\partial x}{\partial y}\Big|_{f,z} = \frac{1}{\frac{\partial y}{\partial x}\Big|_{f,z}} \tag{7.12}$$

$$\frac{\partial y}{\partial x}\Big|_{f,z} = \frac{(\partial y/\partial t)_{f,z}}{(\partial x/\partial t)_{f,z}} \tag{7.13}$$

$$\frac{\partial y}{\partial x}\Big|_{f,z} = -\frac{(\partial f/\partial x)_{y,z}}{(\partial f/\partial y)_{x,z}} \tag{7.14}$$

We now proceed to outline a process for deriving expressions for any second derivative in terms of the three in (7.11).

Rule 1: If the derivative contains any potentials, bring them one by one to the numerator and eliminate them with their differential form.

Example 7.1. $(\partial P/\partial U)_{T,N} =?$

Use (7.12)

$$\frac{\partial P}{\partial U}\Big|_{T,N} = \frac{1}{(\partial U/\partial P)_{T,N}} \tag{7.15}$$

Use $dU = TdS - PdV$ at constant N to get

$$\frac{\partial U}{\partial P}\Big|_{T,N} = T\frac{\partial S}{\partial P}\Big|_{T,N} - P\frac{\partial V}{\partial P}\Big|_{T,N} \tag{7.16}$$

The thermodynamic potential U has been eliminated.

Example 7.2. $(\partial T/\partial P)_{H,N} =?$

Use (7.14)

$$(\partial T/\partial P)_{H,N} = -\frac{(\partial H/\partial P)_{T,N}}{(\partial H/\partial T)_{P,N}} \tag{7.17}$$

Use $dH = TdS + VdP$ at constant N twice

$$(\partial T/\partial P)_{H,N} = -\frac{T(\partial S/\partial P)_{T,N} + V}{T(\partial S/\partial T)_{P,N}} \tag{7.18}$$

All thermodynamic potentials have now been eliminated, but unknown derivatives are still present.

Rule 2: If a derivative contains the chemical potential, bring it to the numerator and eliminate it with the Gibbs-Duhem relation $d\mu = -sdT + vdP$.

Example 7.3.

$$\frac{\partial\mu}{\partial S}\Big|_{T,N} = \frac{V}{N}\frac{\partial P}{\partial S}\Big|_{T,N} \tag{7.19}$$

Rule 3: If a derivative contains the entropy, bring it to the numerator. Check whether you can eliminate it with a Maxwell relation. If not, insert a derivative with respect to temperature and you get a specific heat, either at constant pressure or constant volume.

Example 7.4. $(\partial S/\partial P)_{T,N} =?$

$$\frac{\partial S}{\partial P}\Big|_{T,N} = -\frac{\partial^2 G}{\partial P\partial T} = -\frac{\partial^2 G}{\partial T\partial P} = -\frac{\partial V}{\partial T}\Big|_P = -V\alpha \tag{7.20}$$

Example 7.5. $(\partial S/\partial V)_{P,N} =?$

There is no thermodynamic potential that has V, P, and N as its natural variables; thus there is no Maxwell relation. Therefore we insert a derivative with respect to temperature using (7.13)

$$\frac{\partial S}{\partial V}\Big|_{P,N} = \frac{(\partial S/\partial T)_{P,N}}{(\partial V/\partial T)_{P,N}} = T\frac{V}{N}\frac{c_P}{\alpha} \tag{7.21}$$

Rule 4: If the volume is involved in a derivative, bring it to the numerator and express it in terms of α and κ_T.

Example 7.6. $(\partial T/\partial P)_{V,N} =?$

Use (7.14) to get

$$\frac{\partial T}{\partial P}\Big|_{V,N} = -\frac{(\partial V/\partial P)_{T,N}}{(\partial V/\partial T)_{P,N}} = \frac{\kappa_T}{\alpha} \tag{7.22}$$

At this stage we should have expressed any derivative in terms of α, κ_T, and c_P except that the specific heat at constant volume

$$c_V = \frac{T}{N} \frac{\partial S}{\partial T}|_{V,N} \tag{7.23}$$

might still be appearing. To relate it to the fundamental three we start from the entropy as a function of T, P, and N and write its differential

$$dS = \frac{\partial S}{\partial T}|_{P,N} dT + \frac{\partial S}{\partial P}|_{T,N} dP \tag{7.24}$$

We take a derivative of both sides with respect to T at constant V

$$\frac{\partial S}{\partial T}|_{V,N} = \frac{\partial S}{\partial T}|_{P,N} + \frac{\partial S}{\partial P}|_{T,N} \frac{\partial P}{\partial T}|_{V,N} \tag{7.25}$$

Multiplying with T/N, the left side is c_V, the first term on the right is c_P, and both factors in the second term we did already. Thus we get **Rule 5:**

$$c_V = c_P - T \frac{V}{N} \frac{\alpha^2}{\kappa_T} \tag{7.26}$$

This by itself is an interesting result because it states that heating a system up at constant volume costs less energy and is more efficient as no work is done to change the volume. Using a pressure cooker at fixed volume or simply putting a tight lid on a pot costs less and is faster (and less smelly!) than leaving the lid off to cook under constant pressure conditions.

For an ideal gas we have $\alpha = 1/T$ and $\kappa_T = 1/P$ so that $c_P = c_V + R$.

7.3 Applications

7.3.1 *Adiabatic compression*

Compression of a gas implies an increase in its pressure (except during a phase transition). If this is done adiabatically, i.e. while the pressure chamber is thermally isolated from the outside world, no heat is transferred and, according to the relation $dQ = TdS$, it happens at constant entropy. A useful bit of information might be how much the system heats up in adiabatic compression. Insisting that the process happens under quasi-static conditions, all intermediate states are in equilibrium and we can write

$$dT = \frac{\partial T}{\partial P}|_{S,N} dP \tag{7.27}$$

Using our rules we get

$$\frac{\partial T}{\partial P}\Big|_{S,N} = -\frac{(\partial S/\partial P)_{T,N}}{(\partial S/\partial T)_{P,N}} = T\frac{V}{N}\frac{\alpha}{c_P} \tag{7.28}$$

where we used a Maxwell relation on the numerator and identified the denominator as the specific heat. It will also be of interest to know how the volume changes in adiabatic compression, i.e. we calculate

$$dV = \frac{\partial V}{\partial P}\Big|_{S,N}dP \tag{7.29}$$

The derivative is also a measure of compressibility (isothermal, not adiabatic). Let us therefore define an adiabatic compressibility

$$\kappa_S = -\frac{1}{V}\frac{\partial V}{\partial P}\Big|_{S,N} \tag{7.30}$$

and relate it to the isothermal compressibility.

$$\frac{\partial V}{\partial P}\Big|_{S,N} = -\frac{(\partial S/\partial P)_{V,N}}{(\partial S/\partial V)_{P,N}} \tag{7.31}$$

For both the numerator and the denominator we insert a temperature derivative, for instance

$$\frac{\partial S}{\partial P}\Big|_{V,N} = \frac{\partial S}{\partial T}\Big|_{V,N}\frac{\partial T}{\partial P}\Big|_{V,N} \tag{7.32}$$

so that

$$\frac{\partial V}{\partial P}\Big|_{S,N} = -\frac{(\partial S/\partial T)_{V,N}}{(\partial S/\partial T)_{P,N}}\frac{(\partial T/\partial P)_{V,N}}{(\partial T/\partial V)_{P,N}} \tag{7.33}$$

But, see (7.31),

$$-\frac{(\partial T/\partial P)_{V,N}}{(\partial T/\partial V)_{P,N}} = \frac{\partial V}{\partial P}\Big|_{T,N} \tag{7.34}$$

so that we get

$$\kappa_s = \frac{c_V}{c_P}\kappa_T \tag{7.35}$$

For an ideal gas of atoms we get $\kappa_s = (3/5)/P$ and the volume change in adiabatic compression is given by

$$dV = -\frac{3}{5}VdP \tag{7.36}$$

which, upon integration becomes

$$\ln(V/V_0) = -(3/5)\ln(P/P_0) \tag{7.37}$$

or the adiabat is given by

$$PV^{5/3} = \text{const.} \tag{7.38}$$

Remark 7.1. As we have stressed repeatedly, thermodynamics distills order from the overwhelming amount of data that can be derived from experiment. It makes clear the relationship between different physical processes, and thus can be used to make accurate predictions for a wide range of behavior (i.e. how a system will respond to any given experiment). As the example we just discussed, if we know the expansion coefficient, the specific heat at constant pressure, and the isothermal compressibility - all as functions of temperature and pressure over the relevant range - we can calculate the adiabatic compressibility and, in the next step, the compressional adiabat of a system. No need to do another experiment! Here we demonstrated this for an ideal gas but the formulae used are valid for any system.

Remark 7.2. As an aside: specific heats are measured in calorimeters. What can you do if you do not have one? Well, measure the expansion coefficient, the isothermal compressibility, and the adiabatic compressibility - they are all mechanical measurements - and then calculate the specific heat at constant pressure or at constant volume using (7.26) and (7.35), for any system.

To get the temperature rise for a pressure change from P_i to P_f we integrate

$$T_f = T_i + \frac{1}{N} \int_{P_i}^{P_f} T(P) \frac{\alpha(P, T(P))}{c_P(T, T(P))} dP \tag{7.39}$$

Note that second derivatives are functions of pressure and temperature.

7.3.2 *Isothermal compression*

In isothermal compression we might want to predict what the volume change is for a given pressure change at some temperature. We therefore need to evaluate

$$dV = \frac{\partial V}{\partial P}\Big|_{T,N} dP = -V\kappa_T dP \tag{7.40}$$

That is quite straightforward. A more interesting question is how the entropy changes in isothermal compression

$$dS = \frac{\partial S}{\partial P}\Big|_{T,N} dP = -\frac{\partial V}{\partial T}\Big|_{P,N} dP \tag{7.41}$$
$$= -\alpha V dP$$

where we have used a Maxwell relation from the Gibbs free energy. Thus in isothermal compression there is a heat transfer

$$đQ = TdS = -\alpha V dP \tag{7.42}$$

which we can integrate to get the total heat transfer

$$Q_{i \to f} = - \int_{P_i}^{P_f} \alpha V dP \tag{7.43}$$

where, again, α and V are functions of the pressure (temperature is constant).

7.3.3   Free expansion

We have a gas of N moles enclosed in a volume V_i that is separated from an evacuated volume $V_f - V_i$ by an impermeable wall. At some time we remove the wall by either fracturing it or opening a valve connecting the two volumes. As a result the gas will expand to eventually fill the complete volume V_f. We call this free expansion because we neither do work nor transfer heat. As a result the internal energy of the gas will remain constant. What is the temperature and pressure of the final state?

Because our known variables are U, V, and N we write the temperature as a function of them $T = T(U, V, N)$ and get for the overall change in temperature

$$T_f - T_i = T(U, V_i, N) - T(U, V_f, N) \tag{7.44}$$

and similarly for the pressure. Note that free expansion is irreversible and also not a quasi-static process as the intermediate states are not equilibrium states - thermodynamics cannot make any statements about them. On the other hand, if the expansion is minuscule then we can write

$$dT = \frac{\partial T}{\partial V}\Big|_{U,N} dV \tag{7.45}$$

To evaluate the derivative we start applying **Rule 1**

$$\frac{\partial T}{\partial V}\Big|_{U,N} = -\frac{(\partial U/\partial V)_{T,N}}{(\partial U/\partial T)_{V,N}} \tag{7.46}$$

Continuing with the numerator we use the second part of **Rule 1** inserting $dU = TdS - PdV$ (remember $N = const.$) and get

$$\frac{\partial U}{\partial V}\Big|_{T,N} = T\frac{\partial S}{\partial V}\Big|_{T,N} - P$$

$$= T\frac{\partial P}{\partial T}\Big|_{V,N} - P \tag{7.47}$$

where we used a Maxwell relation from the Helmholtz free energy. For the newly appeared derivative we use **Rule 4** to bring the volume to the numerator

$$\frac{\partial P}{\partial T}\bigg|_{V,N} = -\frac{(\partial V/\partial T)_{P,N}}{(\partial V/\partial P)_{T,N}} = \frac{\alpha}{\kappa_T} \tag{7.48}$$

Similarly we proceed with the denominator of (7.46) and get finally

$$dT = \left(\frac{P}{Nc_V} - \frac{T\alpha}{Nc_V\kappa_T}\right)dV \tag{7.49}$$

The derivation of the pressure change is left as an exercise.

7.3.4 *Joule-Thomson throttling process*

Problem: A non-ideal gas undergoes a throttling process from an initial pressure P_i to a final pressure P_f. The initial temperature and molar volume are T_i and $v_i = V_i/N$. Calculate the final temperature T_f if it has been measured that

$\kappa_T = A/v^2$ and $\alpha = \alpha_0$ along the $T = T_i$ isotherm $(A > 0)$
$c_P = c_P^0$ along the $P = P_i$ isobar.

What is the condition on T_i in order that the temperature be lowered by the expansion?

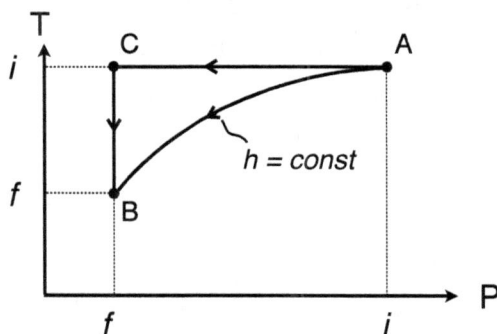

Fig. 7.1   Schematic for the integration of equation 7.52.

Solution: Recall that in a throttling process (i) pressure decreases, (ii) temperature change may be increasing or decreasing, (iii) enthalpy is unchanged for end points, $h_f = h_i$, (iv) if the process is quasi-static then it can be represented by a locus in the $T - P$ plane.

If a process is not quasi-static we can still evaluate along some quasi-static paths with the same end points, in our case for the path $A \to C \to B$ as we know κ_T along the isotherm $A \to C$, and c_P along the isobar $C \to B$, see Figure 7.1.

(1) Along $A \to C$ an infinitesimal change in pressure results in a change in the molar enthalpy

$$dh = \frac{\partial h}{\partial P}\big|_{T_i,N} dP \tag{7.50}$$

Since $dh = Tds + vdP$ we get

$$dh = \left[T\frac{\partial s}{\partial P}\big|_{T_i,N} + v \right] dP \tag{7.51}$$

Using a Maxwell relation from the Gibbs free energy we get

$$dh = v(P)\left[1 - T_i\alpha_0\right] dP \tag{7.52}$$

To integrate this for finite changes we need $v = v(P)$. Now we have been given the fact that along the isotherm

$$\varkappa_T = A/v^2 = -\frac{1}{v}\frac{\partial v}{\partial P}\big|_{T_i,N} \tag{7.53}$$

or

$$\frac{\partial v}{\partial P}\big|_{T_i,N} = -A/v \tag{7.54}$$

which we can integrate to get along the isotherm

$$\frac{1}{2}v^2 = -AP + f(T_i) \tag{7.55}$$

We have now everything to integrate (7.52) along the isotherm. To do this it is advantageous to change the integration variable from dP to dv using (7.55)

$$vdv = -AdP \tag{7.56}$$

to get

$$h_C - h_A = \frac{\alpha_0 T_i - 1}{3A}(v_C^3 - v_A^3) \tag{7.57}$$

(2) Next we look at the path $C \to B$ along which we need

$$dh = \frac{\partial h}{\partial T}\big|_{P_f,N} dT \tag{7.58}$$

$$= T\frac{\partial s}{\partial T}\big|_{P_f,N} dT = c_P^0 dT$$

and upon integration

$$h_B - h_C = c_P^0(T_i - T_f) \tag{7.59}$$

(3) On the other hand, the enthalpy is constant along the $A \to B$ path so that

$$(h_C - h_A) + (h_B - h_C) = h_B - h_A = 0 \qquad (7.60)$$

or

$$\frac{\alpha_0 T_i - 1}{3A}(v_C^3 - v_A^3) = c_P^0(T_i - T_f) \qquad (7.61)$$

To get cooling in the throttling process, $T_f < T_i$, we must have $(v_C^3 - v_A^3) > 0$ which is the case when $(v_C - v_A) > 0$. Along the path $A \to C$ we have from (7.55)

$$A(P_A - P_B) = \frac{1}{2}(v_C^2 - v_A^2) = \frac{1}{2}(v_C - v_A)(v_C + v_A) \qquad (7.62)$$

Now, $P_A = P_i > P_f = P_C$ so that also $v_C > v_A$ and the throttling process leads to a reduction of temperature if

$$\alpha_0 T_i - 1 > 0 \qquad (7.63)$$

This is a nontrivial result because the thermal expansion coefficient is only known along the isotherm $A \to C$ and not along the isenthalp $A \to B$.

7.3.5 *Heating a room*

In 1938, Robert Emden published a short article in Nature [Emden (1938)] with the title "Why do we have Winter Heating?". The objective of this work was to demonstrate the overriding importance of the entropy over the energy. Emden begins by offering two answers to the question posed in his title:

> The layman will answer: 'To make the room warmer'. The student of thermodynamics will perhaps express it thus: 'To import the lacking (internal thermal) energy'. If so, then the layman's answer is right, the scientist's is wrong.

Emden then proves his assertion with a simple argument based solely on the ideal gas for air. It is worth studying this problem with the formal approach developed in this chapter.

What happens thermodynamically when you heat a room to increase its temperature? The quick answer is: you turn on a heater that adds heat thus increasing the internal energy raising the temperature. Although that may be the case for some rooms, it is not so for rooms we live in. Let us analyze this "quick" answer first before we proceed with a realistic analysis.

Heat transferred to a system will be used solely to increase its internal energy only if the walls are rigid (otherwise some of the heat transferred could be used to do work on the outside) and impermeable (otherwise air will escape taking some of the heat with it). Such rooms with rigid and impermeable walls exist but hardly qualify as living rooms, for reasons the reader can easily come up with.

The walls surrounding a room can be considered as rigid because thermal expansion on raising the temperature of the air inside is negligible, so that the volume of the room is constant. We can also assume that the walls are adiabatic except for the surface of the heater, which must be diathermal to transfer heat (we neglect conductive heat loss through doors and windows). The walls of a room are approximately impermeable, except of course the doors and windows. Thus overall the walls must be treated as permeable so air can escape and enter. Permeability implies that the air pressure inside and outside the room is the same and remains constant as anything we do to the room will hardly effect the world outside; at least we hope that, otherwise we heat the whole town and more. Preliminary conclusion: during the process of raising the temperature, pressure and volume of the room remain constant, the amount of air in the room (mole number) does not.

Simplified solution: we treat the air, our thermodynamic system, as an ideal gas so that $PV = NRT$ and, with P and V both constant, the internal energy $U = c_V PV/R = c_V NT$ remains constant. This implies that raising the temperature of the room can only be done by expelling air. The thermodynamic property that changes is the entropy. To simplify the calculation at this point we calculate the change of entropy for fixed V and U rather than $P = UR/(c_V V)$. Thus we need to know

$$dS = \frac{\partial S}{\partial T}|_{U,V} dT$$

$$= \frac{\partial S}{\partial N}|_{U,V} \frac{\partial N}{\partial T}|_{U,V} dT$$

$$= (-\frac{\mu}{T})(-\frac{N}{T}) dT \tag{7.64}$$

But the chemical potential of an ideal gas at room temperature is negative. The result: as the temperature is raised in the room the entropy *decreases* by expelling air to the outside. Thus the heat transfer $dQ = TdS$ is also negative, i.e. we indeed transfer heat to the outside to heat a room. Also note that the energy per mole (or per volume) increases as the temperature increases.

General solution: We repeat the calculations now without resorting to an ideal gas. This is a good exercise in thermodynamic thinking and rigor. We again need the entropy change for a given temperature change but this time at constant pressure and volume

$$dS = \frac{\partial S}{\partial T}\Big|_{P,V}\, dT$$

$$= \frac{\partial S}{\partial N}\Big|_{P,V}\frac{\partial N}{\partial T}\Big|_{P,V}\, dT \tag{7.65}$$

Write

$$\frac{\partial N}{\partial T}\Big|_{P,V} = -\frac{(\partial V/\partial T)_{N,P}}{(\partial V/\partial N)_{T,P}} \tag{7.66}$$

The numerator is the thermal expansion coefficient up to a factor V. For the denominator we have a Maxwell relation

$$\frac{\partial V}{\partial N}\Big|_{T,P} = \frac{\partial \mu}{\partial P}\Big|_{T,N} = \frac{V}{N} \tag{7.67}$$

because for T constant the Gibbs-Duhem equation reads $d\mu = (V/N)dP$. Overall, we thus have

$$\frac{\partial N}{\partial T}\Big|_{P,V} = -N\alpha \tag{7.68}$$

Turning now to the first factor in (7.65) we write

$$\frac{\partial S}{\partial N}\Big|_{P,V} = -\frac{(\partial P/\partial N)_{S,V}}{(\partial P/\partial S)_{N,V}} \tag{7.69}$$

to get, as the arguments, the natural variables of a thermodynamic potential, namely the internal energy. This guarantees the existence of a useful Maxwell relation for the numerator

$$-\frac{\partial P}{\partial N}\Big|_{S,V} = \frac{\partial \mu}{\partial V}\Big|_{S,V}$$

$$= -s\frac{\partial T}{\partial V}\Big|_{S,N} + v\frac{\partial P}{\partial V}\Big|_{S,N} \tag{7.70}$$

For the denominator we have a Maxwell relation

$$\frac{\partial P}{\partial S}\Big|_{N,V} = -\frac{\partial T}{\partial V}\Big|_{S,N} \tag{7.71}$$

so that

$$\frac{\partial S}{\partial N}\Big|_{P,V} = s - v\frac{\partial P}{\partial V}\Big|_{S,N}\frac{\partial V}{\partial T}\Big|_{S,N} = s - v\frac{\partial P}{\partial T}\Big|_{S,N}$$

$$= s - v\frac{\partial S}{\partial T}\Big|_{P,N}\frac{\partial T}{\partial V}\Big|_{P,N}$$

$$= s - \frac{1}{T}\frac{c_P}{\alpha} \tag{7.72}$$

Overall we get for (7.65)

$$\frac{\partial S}{\partial T}\Big|_{P,V} = -N\alpha \left(s - \frac{c_P}{T\alpha}\right) \tag{7.73}$$

Using the Euler relation for molar quantities, $s = u/T + Pv/T - \mu/T$, we can show that this general result agrees with (7.64) for an ideal gas. Further aspects of this problem are formulated in Problem 7.5.

Remark 7.3. This analysis of heating a room again demonstrates the all-important role that the entropy plays in thermodynamic thinking. To make this point memorable, we finish with another quote from the conclusions of Emden's article:

> As a student, I read with advantage a small book by F. Wald entitled "The Mistress of the World and her Shadow". These meant energy and entropy. In the course of advancing knowledge the two seem to me to have exchanged places. In the huge manufactory of natural processes, the principle of entropy occupies the position of manager, for it dictates the manner and method of the whole business, whilst the principle of energy merely does the book-keeping, balancing credits and debits.
> – R. Emden

7.4 Problems

Problem 7.1. *A real gas is well described by the van der Waals equation of state for 1 mole (v=V/N)*

$$(P + \frac{a}{v^2})(v - b) = RT$$

One mole of this gas is expanded isothermally at temperature T from an initial volume v_i to a final volume v_f. Find the heat transfer to the system in this expansion.

Problem 7.2. *One mole of a van der Waals fluid with the fundamental relation*

$$S(U, V, N) = Ns_0 + NR \ln\left[(v - b)(u + a/v)^c\right]$$

is contained in a vessel of volume V_i, at temperature T_i. A valve is opened, permitting the fluid to expand freely into an initially evacuated vessel, so that the final volume is V_f.

(a) The walls of the vessel are adiabatic. Find the final temperature T_f.

(b) Find the final temperature when the expansion is done quasi-statically and adiabatically against a piston.

Problem 7.3. *Two states (A and B) of a mole of a particular system lie on the curve $Pv^2 = const$ such that $P_A v_A^2 = P_B v_B^2 = const$. Along this curve it has also been measured that $c_P = Cv^2$, $\alpha = D/v$ and $\kappa_T = Ev$ with constants C, D, E. Calculate T_B in terms of T_a, P_A, v_A, v_B. Hint: Write the differential form for $T = T(v, P)$ and eliminate dP along the line $Pv^2 - const$.*

Problem 7.4. *A solid sphere of mass $m = 10$ g can slide snugly in a vertical cylindrical tube of cross sectional area $A = 10$ cm^2, connected at the bottom to a gas container of volume $V = 1$ l and temperature $T = 20°$ C. The upper end of the tube is open and exposed to the atmosphere at pressure P.*

(a) What is the pressure P_{eq} in the gas container when the sphere is at rest?

(b) Displacing the sphere and releasing it will start oscillations around the equilibrium position. Calculate the frequency ν of the oscillations for small displacements assuming further that the motion itself is slow enough to be treated as quasi-static and fast enough to suppress heat conduction so that it is adiabatic.

(c) What is v for the parameters given above.

Hint: the sphere displaced by a vertical distance Δx experiences a restoring force $F_{restoring} = -k\Delta x$ resulting in oscillatory motion of angular frequency $\omega = \sqrt{k/m}$. Relate the restoring force to the pressure change induced by the volume change $\Delta V = A\Delta x$.

Problem 7.5. *Heating a room:*

(a) Show that the general result (7.73) agrees with the simpler result for an ideal gas (7.64). Hint: use the entropy for an ideal gas remembering that $(-\mu/T) = \partial S/\partial N_{U,V}$.

(b) Integrate (7.64) for a finite temperature change from T_i to T_f using (4.51) for the chemical potential.

(c) For a hermetically sealed room no gas can escape. Assume that the walls are moveable so that the pressure stays fixed. Calculate the change in

volume ΔV for a change in temperature ΔT. Assume that the temper-ature is raised from freezing to room temperature. How much would the walls of an average room of 4 m $\times$ 5 m $\times 2.5$ m have to move to keep the the pressure constant? Is this reasonable?

(d) *For the same hermetically sealed room we fix the walls so that the pres-sure will rise when heating. Calculate the change in volume ΔP for a change in temperature ΔT. Assume that the temperature is raised from freezing to room temperature. How much will the pressure rise in an average room of 4 m $\times$ 5 m $\times 2.5$ m? Is this reasonable?*

Problem 7.6. *Eliminate c_P from (7.26) using (7.35) to express c_V in terms of α, κ_T, and κ_S.*

Problem 7.7. *Calculate the pressure change $P_f - P_i$ in free expansion of a gas.*

Chapter Summary

(1) The fact that second derivatives of well-behaved functions are symmet-ric in the order of differentiation is used to derive Maxwell relations, a total of 27 for a simple one-component system.
(2) A set of rules is formulated that allows the reduction of any second derivative to the three fundamental quantities c_P, α, and κ_T.
(3) A number of realistic problems such as isothermal expansion, Joule-Thomson throttling, etc. are analyzed.

Chapter 8

Engines, Hurricanes, and Athletes

In an engine, a cyclic process makes energy available to perform a well-defined task. The process must be cyclical so that the same engine can be used over and over again with the engine returning, apart from wear and tear, to its initial state after each cycle is complete. Our task in this chapter will be to derive general criteria about the feasibility and efficiency of engines. The guiding principles will be the first and second law of thermodynamics.

According to the first law of thermodynamics energy is conserved, implying that energy cannot be created out of nothing. Throughout history, many machines have been proposed that supposedly overcome this law. Such an engine is frequently called a perpetual motion machine of the first kind and, according to the first law, such a machine is not possible. What is allowed under the first law, however, is the conversion of one type of energy to another. An example is a sailboat which converts the kinetic energy of the wind into the mechanical motion of the boat. This conversion is not perfect, however, as part of the energy is dissipated as friction with the water, which in turn creates waves and heats up the water; air turbulence also reduces the efficiency. Another example is an electrical engine such as a kitchen blender which converts the kinetic energy of moving electrons (i.e. the current) into mechanical energy to cut up vegetables. Again there are side effects, namely friction in the electrical wire leading to Ohmic heating, and friction in the mechanical motion within the blender.

The first law of thermodynamics also allows for the conversion of heat into mechanical energy, because it says that at constant mole number

$$dU = TdS - PdV \qquad (8.1)$$

We have stated previously that energy in the form of heat is not necessarily available to do work. An example of such a machine would be a ship that extracts heat from the ocean and converts it into propulsion. The resulting cooling of the ocean would be minimal and also offset by the fact that the motion of the ship puts some of this heat back through friction. Yet, overall the entropy of the system - ocean plus ship - would be decreased in contradiction to the second law of thermodynamics. Thus such a machine is not possible. To make this point more precise we formulate the second law as follows:

Claim 8.1. *It is impossible to construct a cyclic engine in which the only change after one cycle is the extraction of heat from one reservoir for the purpose of performing work. Such an engine is called a perpetual motion machine of the second kind; it is impossible according to the second law.*

Consider an engine design where there is only one heat reservoir available. In the isothermal expansion of an ideal gas the energy, $U = cNRT$, remains constant, i.e. the work done is equal to the heat extracted from the heat reservoir that keeps the temperature constant. If you were to make this a cyclic process you must follow up the isothermal expansion by an isothermal compression in which the heat extracted in the expansion phase will be returned and no work is done!

Thus according to the second law of thermodynamics a cyclical engine can only deliver work by extracting heat from a high temperature heat reservoir if a second, colder reservoir is available into which heat, and thus entropy, is transferred at a rate such that overall the entropy increases. In this statement no mention is made on the nature of the heat reservoirs or the engine; how the engine is constructed is left to the ingenuity of the inventor. No matter how clever the inventor, the second law puts a limit on the maximum efficiency that can be achieved and depends solely on the temperatures of the two heat reservoirs. This limit was derived by Sadi Carnot long before the principles of thermodynamics were firmly established.

8.1   The Carnot cycle

Carnot based his arguments on the ideal gas. Although his result for engine efficiency is valid for any system, it is instructive to first follow his simple arguments.

We start with an ideal gas in a container fitted with a moveable piston so that in the initial state it occupies a volume V_A and is in thermal contact

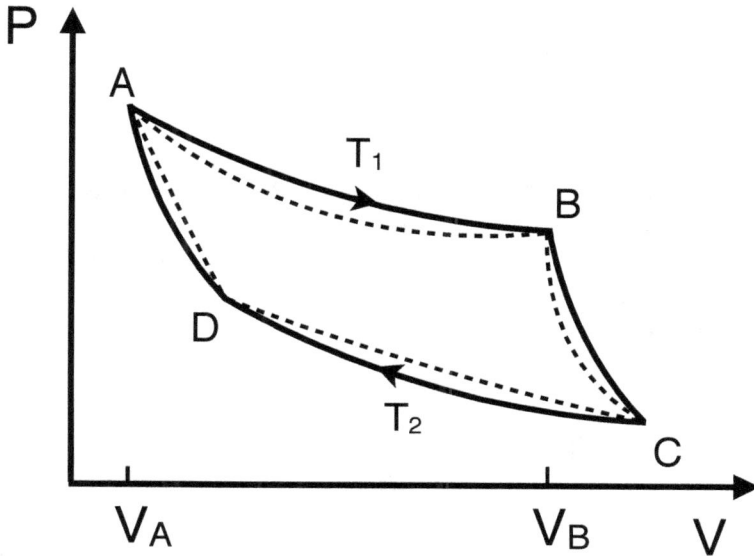

Fig. 8.1 Carnot cycle in the $P - V$ plane. $A \to B$ and $C \to D$ are isotherm processes; $B \to C$ and $D \to A$ are adiabatic. Dashed lines indicate a realistic duty cycle.

with a heat reservoir at a temperature T_h. Its pressure is thus $P_A = NRT_h/V_A$. We now perform four processes, see Figure 8.1:

(1) Keeping the gas in contact with the heat reservoir at T_h we expand it isothermally to a volume V_B so that its pressure drops to $P_B = NRT_h/V_B$.

(2) We remove the thermal contact and expand the gas adiabatically to a volume V_C. Along the adiabat we have for an atomic gas $TV^{2/3} = const$, so that the gas cools to a temperature $T_c = (V_C/V_B)^{2/3}T_h$.

(3) Next we connect the gas thermally to a heat reservoir at the colder temperature T_c and compress it isothermally to a volume V_D so that its pressure is $P_D = NRT_c/V_C$.

(4) Lastly, we disconnect the thermal contact to T_c and adiabatically compress the gas to the initial state (V_A, T_h).

Let us look at the energy balance after a complete cycle: the change in internal energy is zero so that the work delivered is given by $W = Q_{A \to B} - Q_{C \to D}$, because no heat is exchanged in the two adiabatic processes

$A \to B$ and $C \to D$. For the heat transfer in the isothermal expansion and compression we have $Q_{A \to B}/T_h = Q_{C \to D}/T_c$, so we get for the complete cycle

$$W_{cycle} = Q_{A \to B} \frac{T_h - T_c}{T_h} \tag{8.2}$$

We are looking now for an expression to quantify how much heat must be made available from the hot reservoir to an engine in order to accomplish a certain amount of work. This heat might be generated by burning fuel. We call this ratio the cycle efficiency

$$\varepsilon = \frac{W_{cycle}}{Q_{A \to B}} \tag{8.3}$$

For the Carnot cycle this turns out to be

$$\varepsilon = \frac{T_h - T_c}{T_h} \tag{8.4}$$

where ε is the fraction of the heat extracted from the hot reservoir that is converted into work. The rest of the extracted heat $Q_{C \to D} = Q_{A \to B}(T_c/T_h)$ is dumped into the cold reservoir. Unless we have access to an even colder reservoir, this heat is no longer accessible to us.

If $T_h = T_c$ we have only one heat reservoir and no work can be extracted. To increase the efficiency to 1, we need a cold reservoir that is much colder. Also note that using thermodynamics we assumed that all processes are done quasi-statically, i.e. very slowly. To transfer heat from the hot reservoir at T_h in a finite time, the gas must be somewhat colder than T_h, and similarly hotter than T_c, as indicated by the dashed line in Figure 8.1. From the point of view of thermodynamics, the faster an engine is run (making the process less quasi-static), the lower the efficiency will be. Also note that $1 - \varepsilon$ is the fractional amount of heat that has to be transferred to the cold reservoir as thermal pollution.

8.2   Maximum work theorem

To derive a general statement about the maximum efficiency at which a thermodynamic engine can operate, we consider a composite system. The primary system is coupled to both a reversible work source (some mechanical machine) with which it can only exchange work, and a reversible heat source with which it can only exchange heat. This is shown in Figure 8.2.

A reversible work source is a system that is enclosed by adiabatic, impermeable walls (no heat exchange) but with movable walls to do work. Conversely, a reversible heat source has rigid but diathermal, impermeable walls so that heat can be exchanged. With the attribute 'reversible' we emphasize that heat and work can go in either direction, i.e. the 'sources' can also be sinks.

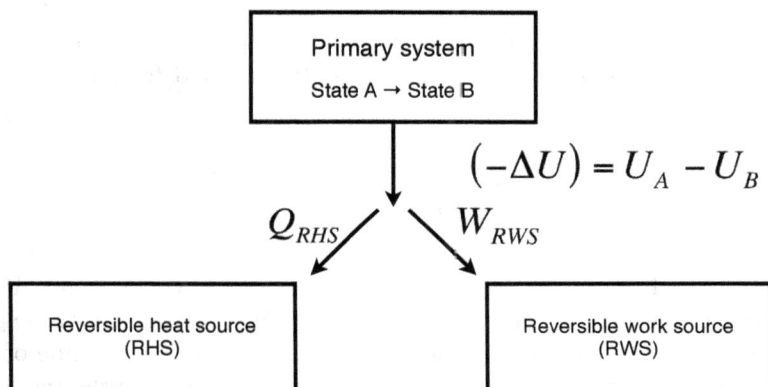

Fig. 8.2   Schematic for the derivation of the maximum work theorem.

We take the primary system from a state A to a state B by extracting an amount of internal energy, $(-\Delta U) = U_A - U_B$. Part of this internal energy is used to perform work ΔW_{RWS} with the reversible work source, and part is delivered to the colder reversible heat source as heat ΔQ_{RHS}. Treating these processes as quasi-static (small and slow compared to the internal relaxation time) energy conservation demands

$$dU + dW_{RWS} + dQ_{RHS} = 0 \tag{8.5}$$

while the entropy maximum principle requires

$$dS_{total} = dS + \frac{dQ_{RHS}}{T_c} \geq 0 \tag{8.6}$$

where dS is the entropy change in the primary system. It follows that

$$dW_{RWS} \leq T_c dS - dU \tag{8.7}$$

and the equality holds for reversible processes for which the available work is maximal. To continue we write $dU = dQ + dW$ where $dQ = TdS$ is the heat extracted from the primary system and dW is the work done by it directly.

$$dW_{RWS}(maximum) = \frac{T_c}{T}dQ - dU$$

$$= \left[1 - \frac{T_c}{T}\right](-dQ) + (-dW_{direct}) \qquad (8.8)$$

That is, in a quasi-static process, the maximum work that can be delivered to a work source is the sum of

(1) the work $(-dW)$ directly extracted from the primary system,
(2) a fraction $(1 - T_c/T)$ of the heat $(-dQ)$ extracted from the primary system at temperature T.

The fraction of extracted heat that can be converted to work,

$$\varepsilon = \frac{dW_{RWS}}{(-dQ)} = 1 - \frac{T_c}{T} \qquad (8.9)$$

is the thermodynamic efficiency in accordance with Carnot's result. If the primary system is at the same temperature as the heat reservoir, the engine efficiency is zero. This corresponds to a perpetual motion machine of the first kind. To increase the engine efficiency we have to lower the temperature of the reversible heat source or raise the temperature of the primary system.

Let us look at a simple example taking for the primary system a boiler of hot water, $T = 373$ K, and as the heat reservoir the ambient atmosphere, $T_c = 300$ K. The maximum efficiency achievable with perfect engineering is only 20%! The rest of the extracted heat, 80% of it, must be dumped into the environment to increase the overall entropy in this process. With this low efficiency, it seems difficult to justify the historic success and ubiquity of steam engines. Is there a way to improve this efficiency? Here comes the ingenuity: build a pressurized boiler to increase the temperature of the steam to, say 300° C and the efficiency goes up to 55%.

So far we have derived the maximum work and the thermodynamic efficiency for an infinitesimal quasi-static process. In a realistic machine, finite changes are achieved for which we have to integrate (8.8). This can be tricky if the reversible heat source is finite because then its temperature will rise as heat is transferred into it. However, in most practical situations the reversible heat source is so large that its temperature does not change appreciably, i.e. it should be a heat reservoir. In such a situation, for a finite cycle we get that

$$W_{RWS}(maximum) = T_{res}\Delta S - \Delta U \qquad (8.10)$$

8.3   Engine efficiency

In an engine an amount of heat dQ is extracted from an energy source at high temperature T_h. A fraction $đW = (1 - T_c/T_h)(-đQ)$ is converted to work to operate some machine and the remaining fraction $T_c/T_h(-đQ)$ is dumped into a cooling system, i.e. a cold heat reservoir at T_c. A schematic is shown in Figure 8.3. We will specify the details for a number of well-known engines.

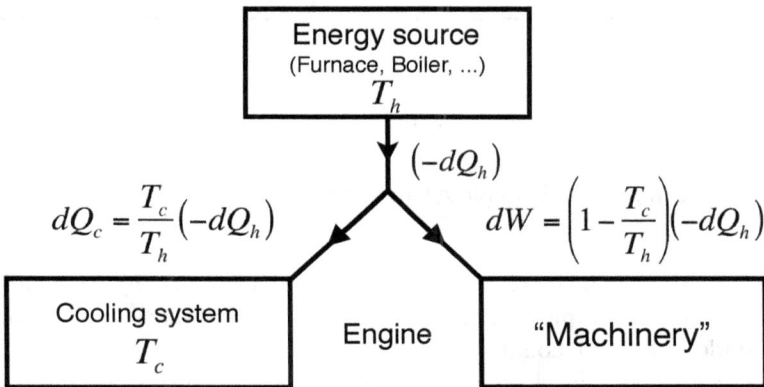

Fig. 8.3   Schematic of a thermodynamic engine.

8.3.1   *Gasoline engine: the Otto cycle*

Remark 8.1 (history). *The gasoline engine was invented by Nikolaus Otto in 1862. It was an indirect-acting free-piston compressionless engine whose great efficiency quickly won most of the market, which at that time was mostly for small stationary engines fueled by lighting gas. In 1876 Otto, working with Gottlieb Daimler and Wilhelm Maybach, developed a practical four-stroke cycle engine, the Otto cycle. In 1879 Karl Benz, working independently, was granted a patent for his internal combustion engine, a reliable two-stroke gas engine, based on the same technology as Nikolaus Otto's design of the four-stroke engine. Later, Benz designed and built his own four-stroke engine that was used in his automobiles, which were developed in 1885, patented in 1886, and became the first automobiles in production, the beginning of the Daimler-Mercedes-Benz tradition.*

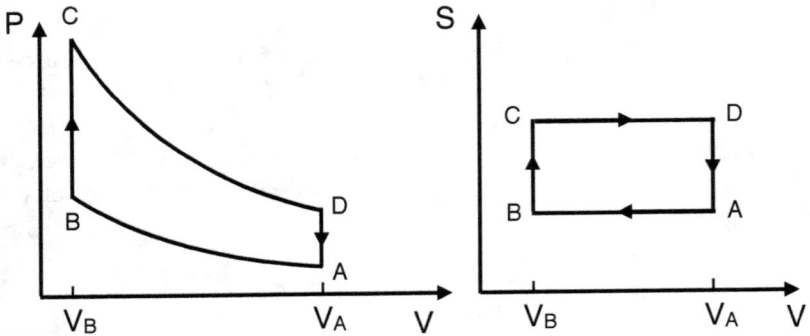

Fig. 8.4   Working diagrams of the Otto cycle.

The working cycle is depicted in Figure 8.4 in the $P - V$ and $S - V$ planes:

A → B : At A a mixture of gasoline vapor and air is injected into the cylinder and rapidly compressed by the upward motion of the piston under adiabatic conditions.

B → C : At B the spark plug is fired to ignite the compressed mixture to lead to combustion and rapid heating at constant volume.

C → D : The increased pressure leads to adiabatic expansion with the piston being driven down in the power stroke.

D → A : At D, the exhaust valve is opened and the hot combustion products (CO, CO_2, etc.) are ejected, together with heat, to the outside which acts as the cold heat reservoir.

Let us approximate the combustible gases as ideal with internal energy $U = cNRT$ where $c = c_V/R > 3/2$ (the gases are certainly not monatomic).

A → B : $Q_{A \to B} = 0$ because this stroke is done adiabatically so that $W_{A \to B} = U_B - U_A = c(P_B V_B - P_A V_A)$. The pressures are connected by the condition of adiabaticity

$$P_B V_B^{c_P/c_V} = P_A V_A^{c_P/c_V} \tag{8.11}$$

B → C : At constant volume no work is done and $Q_{B \to C} = U_C - U_B = c(P_C - P_B)V_B$.

C → D : This is similar to $A \to B$, no heat transfer but adiabatic work $W_{C \to D} = U_D - U_C = c(P_D V_B - P_C V_A)$, and

$$P_D V_A^{c_P/c_V} = P_C V_B^{c_P/c_V} \qquad (8.12)$$

D → A : No work is done and $Q_{D \to A} = U_A - U_C = c(P_D - P_A)V_A$.

Collecting the various contributions to the total work done in the cycle ($W_{A \to B}$ and $W_{C \to D}$) and the heat extracted from the hot reservoir $Q_{B \to C}$ we get the efficiency of the Otto cycle

$$\varepsilon_{Otto} = 1 - \left(\frac{V_B}{V_A}\right)^{\gamma - 1}$$

$$= 1 - \frac{1}{r^{\gamma - 1}} \qquad (8.13)$$

where we recall $\gamma = c_P/c_V$ is the ratio of the specific heats and $r = V_A/V_B$ is the compression ratio. In car engines the compression ratio is typically less than 10 to avoid pre-ignition (remember adiabatic compression heats up a gas!). As late as the 1970's pre-ignition was suppressed by leaded gasoline. In modern multivalve engines with fuel injection, compression and ignition are monitored electronically so one can go beyond that value. For $r = 9$ and $\gamma = 7/5$ (for a diatomic gas) we get $\varepsilon_{Otto} = 0.58$. We can use this number to estimate the ignition temperature from $\varepsilon = 1 - T_c/T_h$ and get roughly $T_h \sim 3T_c$, i.e. for $T_c = 300$ K (ambient air) we get $T_h = 900$ K, which is about right for the combustion of the gasoline vapor-air mixture. In supercharged engines one can achieve temperatures higher than 1000 K. The exhaust fumes are then also quite hot and the engine block needs a lot of cooling so the overall gain in efficiency is not astronomical. However, such engines perform superbly in short spurts as required in car racing but they do not last long.

The overall efficiency of a working gasoline engine is of course much lower than its thermodynamic maximal efficiency because of all the additional power consuming devices such as a cooling system, internal friction, coupling to a transmission, road friction, and more.

8.3.2 *Diesel engine*

After Otto's invention was adopted in the first cars, a lot of effort went into attempts to improve its efficiency by finding ways to avoid pre-ignition or engine knock when higher compression ratios were used. Then in the 1880's came Rudolf Diesel who argued that if you cannot avoid pre-ignition then use it to your advantage. The result was the Diesel engine. Its cycle diagram is shown in Figure 8.5 in the $P - V$ and $T - S$ planes.

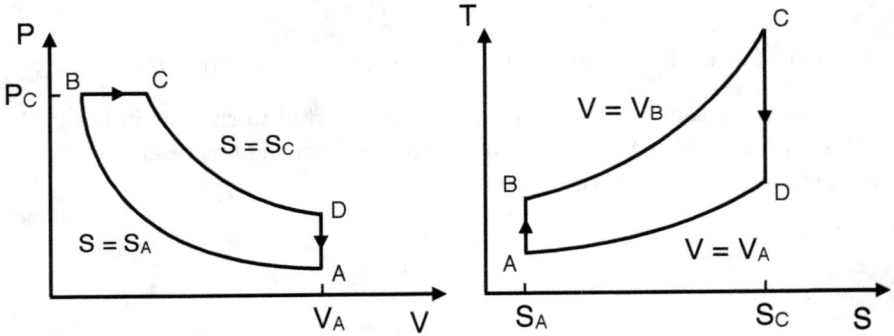

Fig. 8.5   Working diagrams of the Diesel cycle.

$A \rightarrow B$: The piston moves upward, compressing air in the cylinder chamber adiabatically. This increases the temperature but does not lead to pre-ignition because there is no fuel present.

$B \rightarrow C$: At B Diesel fuel is sprayed into the cylinder and ignites spontaneously in the hot air without the need for a spark plug leading to combustion and expansion at constant pressure. If gasoline were sprayed in it would lead to a violent explosion because gasoline is more combustible than Diesel fuel, i.e. pre-ignition would certainly occur. Thus Diesel chose a less volatile fuel, namely Diesel fuel.

$C \rightarrow D$: Expansion continues at a slower pace and becomes adiabatic.

$D \rightarrow A$: At D the exhaust valve is opened and the burnt gas is ejected from the cylinder.

For the calculation of the efficiency of the Diesel engine we again use an ideal gas.

$A \rightarrow B$: $Q_{A \rightarrow B} = 0$ because this stroke is done adiabatically so that $W_{A \rightarrow B} = U_B - U_A = c(P_B V_B - P_A V_A)$. The pressures are connected by the condition of adiabaticity

$$P_B V_B^{c_P/c_V} = P_A V_A^{c_P/c_V} \qquad (8.14)$$

$B \rightarrow C$: Work is done at constant pressure

$$W_{B \rightarrow C} = -\int_{V_B}^{V_C} P_B dV = -P_B(V_C - V_B) \qquad (8.15)$$

and heat is transferred into the system through the combustion

$$Q_{B \rightarrow C} = U_C - U_B - W_{B \rightarrow C} = (c-1)P_B(V_C - V_B) \qquad (8.16)$$

C → D : This is similar to $A \to B$, no heat transfer but adiabatic work $W_{C \to D} = U_D - U_C = c(P_D V_A - P_B V_C)$, and

$$P_D V_A^{c_P/c_V} = P_C V_B^{c_P/c_V} \tag{8.17}$$

D → A : No work is done and $Q_{D \to A} = U_A - U_C = c(P_A - P_D)V_A$.

Collecting the various contributions to the total work done in the cycle $(W_{A \to B} + W_{B \to C} + W_{C \to D})$ and the heat extracted from the hot reservoir $Q_{B \to C}$ we get the efficiency of the Diesel cycle

$$\varepsilon_{Diesel} = 1 - \frac{1}{r^{\gamma - 1}} \frac{r_c^\gamma - 1}{\gamma(r_c - 1)} \tag{8.18}$$

Here $r_c = V_A/V_C$ is the cut-off ratio and $r = V_A/V_B$ the compression ratio. For $r = 9$ and $r_c = 2$ the efficiency of the Diesel engine is $\varepsilon_{Diesel} = 0.51$ as compared to $\varepsilon_{Otto} = 0.58$ for the Otto engine. However, because there is no problem with pre-ignition in the Diesel cycle the compression ratio can be increased to as much as $r = 20$ leading to an efficiency of $\varepsilon_{Diesel} = 0.65$.

8.3.3 *Otto versus Diesel*

There are several technical differences between the Otto and Diesel engine. In the Otto engine, combustion occurs at constant volume, while in the Diesel it is at constant pressure. Control of combustion in the Otto engine is affected by varying the fuel-air mixture. In the Diesel engine this is achieved by varying the amount of fuel injected at constant mass of air. The most important difference, however, is the temperature at which Diesels can operate. Since the Diesel engine has a higher operating temperature than the Otto engine, Diesels operate with higher thermodynamic efficiency. This efficiently results in better fuel economy and lower emissions (with the exception of nitrous oxides).

We note as an aside that although Diesel fuel droplets injected into the combustion chamber burn in about a millisecond, our quasi-static treatment of engine efficiency is still valid and accurate. A process that appears to occur "quickly" may in fact be quasi-static. Recall that the definition of "slowly" is given in terms of characteristic timescales of the system, not the observer.

Because of their higher operating temperature, Diesel engines are built to withstand higher stresses than Otto engines. Consequently, they are more rugged and can produce more torque than an Otto engine of the same size. Higher operating temperatures also allow for the use of less volatile fuels. Diesel fuel does not easily ignite at room temperature. When firefighters train to fight fuel fires, they start diesel fires by first igniting gasoline. Only when the diesel fuel is warm enough will it start to burn!

Unlike gasoline, once ignited, Diesel fuel burns slowly. This has resulted in its widespread adoption in military and industrial applications; Diesel fuel is inherently safer. Diesel engines are also able to accept a larger variety of fuel types than their Otto counterparts. Fuel used in an Otto engine must be sufficiently volatile to allow for vaporization (so the optimal fuel-air mix can be achieved) yet not too volatile, as this will result in pre-ignition. This imposes relatively strict limits on the type of fuel that can be used. Conversely, Diesel engines can run on fuel as simple as waste cooking oil.

The possibility of bio-diesel as a renewable fuel source offers great potential and is an area of active research and development. Another beneficial property of diesel fuel is that it tends to have better lubricating properties than gasoline. This reduces engine wear, thereby increasing operational lifetime. The spark based ignition in Otto engines also necessitates that their design be more complex than Diesels. This results in reduced reliability and higher maintenance costs.

Why then does it seem that Otto engines are more common than Diesels? While this is becoming less and less true (Diesels accounted for more than 50% of new car sales in Europe by the late 2000's), in North America especially there has been a perception that Diesel engines do not perform as well as Otto engines. This view is largely based on some of the historical, and now solved, disadvantages of Diesel engines. Early Diesel engines were notoriously difficult to start in cold weather. This was due to the partial solidification of primitive fuel and the low temperature of the engine block absorbing heat produced during compression. Cold starts are no longer an issue due to the inclusion fuel additives and computer controlled heating elements. Additionally, Otto engines tended to outperform Diesels when bursts of power were needed, i.e. they seemed to have more "pick up". The widespread use of turbo- or super-chargers has reversed this trend, with Diesel engines now offering superior performance. This performance gain is achieved by recycling exhaust fumes for the next compression achieving higher temperatures in the initial adiabatic compression phase. A well-tuned Diesel engine can produce 100 HP at only 750 RPM, whereas an Otto engine would require several thousand RPM for equivalent power production. Another common complaint was that Diesel engines tended to be louder and sometimes produced thick black smoke (due to incomplete combustion). Improvements in engine design and electronic fuel injection have solved these issues. It is likely that as the public perception of Diesel engines continues to improve, their adoption will become increasingly widespread.

The overall efficiency of commercial cars was at the 15% level in the 1970's; in 2000 this had risen to 21%, still a far cry from thermodynamic efficiency.

8.3.4 Refrigerator

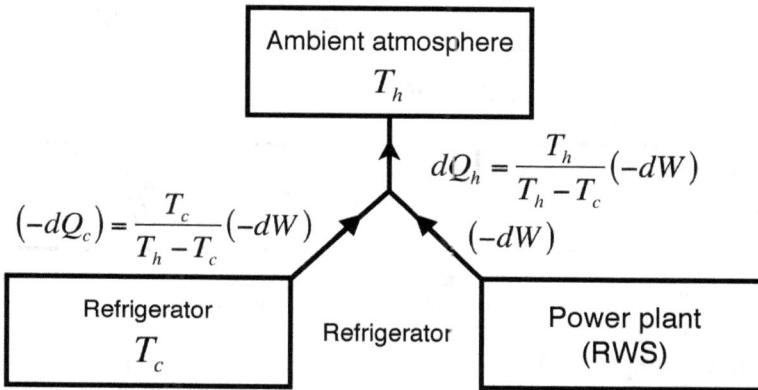

Fig. 8.6   Schematic of a thermodynamic refrigerator.

A refrigerator is a thermodynamic engine working in reverse. Its purpose is to extract heat from the cold system with the input of minimal work, and to eject that heat into the ambient atmosphere, see Figure 8.6. Note that a refrigerator is also a heating device! The efficiency of a refrigerator is given by the coefficient of refrigerator performance - the ratio of the heat removed from the refrigerator to the work that must be purchased from the power company

$$\varepsilon_r = \frac{(-dQ_c)}{(-dW_{RWS})} = \frac{T_c}{T_h - T_c} \tag{8.19}$$

As $T_h \to T_c$, $\varepsilon_r \to \infty$ because no work is needed to transfer heat between two reservoirs at the same temperature. On the other hand, as the interior of the fridge becomes colder the performance becomes smaller. Eventually, as T_c approaches absolute zero the performance goes to zero. The quest for absolute zero becomes increasingly complex (and expensive) and eventually impossible according to Nernst's theorem.

8.3.5   *Heat pump*

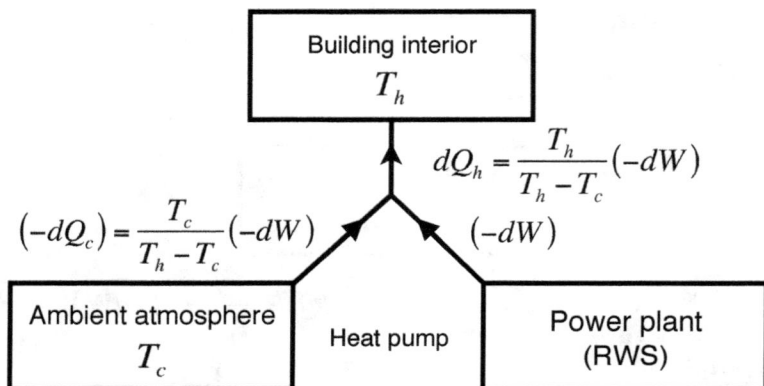

Fig. 8.7   Schematic of a thermodynamic heat pump.

A heat pump is a thermodynamic engine that extracts heat from the colder ambient atmosphere or a large body of water to heat the inside of a house at the cost of doing work. A simple device would be to take the door off a refrigerator and shove it into a door opening: it will try to cool the outside and heat the house in turn. Heat from the heat pump comes from two sources; the cold reservoir and the motor. The coefficient of heat pump performance is the ratio of the heat delivered to the hot system to the power purchased from the power company

$$\varepsilon_p = \frac{dQ_h}{(-dW_{RWS})} = \frac{T_h}{T_h - T_c} \tag{8.20}$$

The colder the outside air the lower the heat pump efficiency.

Example 8.1. To heat a house we can purchase energy from the power company and convert it directly into heat with an electric heater at an efficiency of almost one. On the other hand, we can use this work to run a heat pump. Take the temperature of the inside of the house to be a comfortable $20°$ C $= 293$ K. When the outside temperature is just below that, say $18°$ C, the heat pump performance, i.e. the gain to use a heat pump over direct heating is a factor $\varepsilon_p = 146$. On the other hand, on a cold winter day with $T_c = -20°$ C the gain is reduced to $\varepsilon_r = 7$. So why does not every house run a heat pump? Unfortunately, commercial heat pumps do not achieve the maximum thermodynamic efficiency, quite far from it. A commercial heat pump consists of an outdoor unit with a large fan to blow air over a condenser, a compressor and pump to get the hot heating fluid to the unit inside the house, another fan that blows the air over heating coils and through the heat ducts to the rooms. All this costs energy, reducing the overall efficiency greatly. Below $-20°$ C their efficiency drops to zero. There are too many moving parts and heat losses along the transmitting lines. The performance of an air source heat pump is measured by HSPF (Heating Seasonal Performance Factor). It is the ratio of BTU (-1055 Wsec) heat output over the heating season to Watt-hours of electricity used. An HSPF above 9.0 is considered high efficiency, achieved by most commercial heat pumps. This corresponds to $\varepsilon_P \approx 3$. Considering the initial cost of installing head pumps, this moderate gain in heating efficiency can be difficult to justify for domestic use.

8.4 Cyclones

In meteorology, a cyclone is an area of closed, circular fluid motion rotating in the same direction as the Earth. This is usually characterized by inward spiraling winds that rotate counter clockwise in the Northern Hemisphere and clockwise in the Southern Hemisphere of the Earth. Large-scale cyclonic circulations are almost always centered on areas of low atmospheric pressure. Cyclones have also been seen on other planets outside of the Earth, such as Mars and Neptune. The largest low-pressure systems are cold-core polar cyclones at the polar caps and extratropical cyclones. Warm-core cyclones such as tropical cyclones originate in the tropics. They are also called hurricanes in the Atlantic and typhoons when they originate in the northwestern Pacific.

The climate in the tropics is generally characterized by small daily and seasonal temperature variations. Rainfall is predictable except for local short-lived thunderstorms. Despite this, the tropics breed the worst storm systems on earth - Cyclones. The fact that a cyclone is fed by hot tropical air and then converts this heat into work in the form of devastating winds, thunderstorms and tornadoes suggests that a cyclone acts as a huge engine that one might understand with the help of thermodynamics. The following account is taken from *Physics Today*, and was written by Prof.

Fig. 8.8 A pair of hurricanes, Helene (left) and Gordon (right), as seen by MODIS on NASA's Terra satellite during the 2006 hurricane season. Both storms were rated Category Three at the time this image was taken.

Kerry Emanuel of the Department of Earth, Atmospheric, and Planetary Sciences at MIT.

Reprinted with permission from "Hurricanes: Tempests in a Greenhouse", Physics Today, August 2006. Copyright 2006, American Institute of Physics.

8.4.1   *The greenhouse effect*

Of the solar energy that streams to Earth, about 30% is reflected by clouds or the surface. The radiation that escapes reflection or absorption in the atmosphere is absorbed by the surface, which transmits energy upward both by radiation and in vast convective currents whose visible manifestations are the beautiful cumulus and cumulonimbus clouds that ply the tropical skyscape. The outgoing photons have much longer wavelengths than the incoming photons, since Earth's surface temperature is far lower than the Sun's. The outgoing IR radiation is strongly absorbed by clouds and by trace amounts of certain gaseous components of the atmosphere, notably water vapor, carbon dioxide, and methane. Those constituents reradiate both upward and downward. Remarkably, the surface receives on average more radiation from the atmosphere and clouds than direct radiation from the Sun. The warming of the surface by back radiation from the atmosphere is the greenhouse effect. Because of it, Earth's surface temperature is some 35 K higher than its effective blackbody temperature. That difference makes hurricanes possible.

The relatively high surface temperature also means that atmospheric radiation exports entropy to space. The reason is that the atmosphere is heated at approximately the surface temperature, but it cools at the much lower effective emission temperature of Earth. In equilibrium, the planet must generate entropy, and the vast majority of that entropy is produced in the atmosphere, mainly through the mixing of the moist air inside clouds with the dry air outside them and through frictional dissipation by falling raindrops and snowflakes. Were it not for moisture in the atmosphere, the entropy would have to be produced by frictional dissipation of the kinetic energy of wind. The resulting air motion would be too violent to permit air travel.

Water in the atmosphere thus has a paradoxical effect on climate. It is far and away the most important greenhouse molecule in the atmosphere and is responsible for a surface temperature increase that requires the production of entropy. On the other hand, mixing and irreversible processes associated with precipitation absorb most of the entropy production and spare people from violent winds. But not always.

8.4.2 A Carnot engine

In the part of the tropics where the sea surface is warm enough and the projection of Earth's angular velocity vector onto the local vertical axis is large enough, random small-scale convective currents sometimes organize into rotating vortices known as tropical cyclones. In computer models of the tropical atmosphere, such organization can happen spontaneously, but usually only if a combination of ocean temperature and rotation is somewhat higher than those observed in nature. In subcritical conditions, some trigger is necessary to initiate the vortices, and in the terrestrial atmosphere tropical cyclones only develop from preexisting disturbances of independent origin. In mathematical parlance, tropical cyclones may be said to result from a subcritical bifurcation of the radiative–convective equilibrium state. About 10% of them develop in the Atlantic Ocean, where the disturbance is often a 100-km-scale "easterly wave" that forms over sub-Saharan Africa

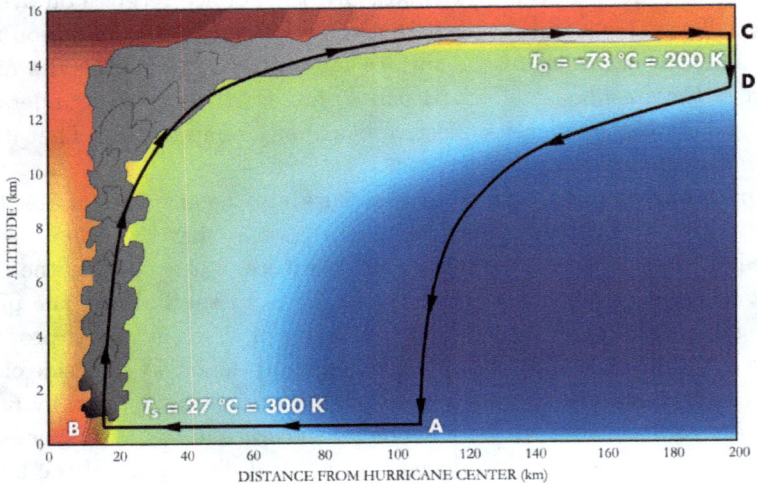

Fig. 8.9   The hurricane as a Carnot heat engine. The plot shows the vertical cross section of the hurricane with its center along the right edge. Colors depict the entropy distribution with cooler colors for lower entropy. The process mostly responsible for a hurricane is the evaporation of sea water that transfers energy from the sea to the air. As a result air spirals inward from A to B and acquires entropy at constant temperature. It then expands adiabatically as it ascends from B to C. Far from the storm center, C to D, it exports IR radiation into outer space losing entropy in a very nearly isothermal compression. This is followed by an adiabatic compression back to A finishing the Carnot cycle.

and then moves westward out over the Atlantic. When its maximum wind speed exceeds 32 m/s, it, by definition, becomes a hurricane.

The convective core of a tropical cyclone may be many tens to hundreds of kilometers across, orders of magnitude greater than the few hundred meters' width of an ordinary cumulus cloud. The core's small surface-to-volume ratio, together with the strong stability to horizontal displacement afforded by the inertial stability of its rotation, greatly reduces mixing between cloudy moist air and clear dry air. In a strong tropical cyclone, entropy production by the mixing of dry and moist air is virtually shut down, and dissipation of the wind's kinetic energy takes over as the primary mechanism for producing entropy. Most of the dissipation occurs in a turbulent atmospheric boundary layer within a few hundred meters of the ocean surface.

The mature hurricane is an almost perfect example of a Carnot heat engine whose working fluid may be taken as a mixture of dry air, water vapor, and suspended condensed water, all in thermodynamic equilibrium. The engine is powered by the heat flow that is possible because the tropical ocean and atmosphere are not in thermal equilibrium. This disequilibrium arises because, thanks to the greenhouse effect, the ocean must lose heat by direct, non-radiative transfer to the atmosphere to balance the absorption of solar radiation and back radiation from the atmosphere and clouds. The heat transfer is accomplished mostly by evaporation of water, which has a large heat of vaporization. To maintain substantial evaporation rates, the air a short distance above the sea surface must be much drier than would be the case were it in equilibrium with the sea.

Figure 8.9 illustrates the four legs of a hurricane Carnot cycle. From A to B, air undergoes nearly isothermal expansion as it flows toward the lower pressure of the storm center while in contact with the surface of the ocean, a giant heat reservoir. As air spirals in near the surface, conservation of angular momentum causes the air to rotate faster about the storm's axis. Evaporation of seawater transfers energy from the sea to the air and increases the air's entropy.

Once the air reaches the point where the surface wind is strongest— typically 5–100 km from the center of the hurricane—it turns abruptly (point B in the figure) and flows upward within the sloping ring of cumulonimbus cloud known as the eyewall. The ascent is nearly adiabatic. In real storms the air flows out at the top of its trajectory (point C in the figure) and is incorporated into other weather systems; in idealized models one can close the cycle by allowing the heat acquired from the sea surface to be isothermally radiated to space as infrared radiation from the storm outflow. Finally, the cycle is completed as air undergoes adiabatic compression from D to A.

The rate of heat transfer from the ocean to the atmosphere varies as vE, where v is the surface wind speed and E quantifies the thermodynamic disequilibrium between the ocean and atmosphere. But there is another source of heat; the dissipation of the kinetic energy of the wind by surface friction. That can be shown to vary as v^3. According to Carnot, the power generation by the hurricane heat engine is given by the rate of heat input multiplied by the thermodynamic efficiency.

If the storm is in a steady condition, then the power generation must equal the dissipation, which is proportional to v^3. Equating dissipation and generation yields an expression for the wind speed:

$$v^2 = \frac{T_s - T_o}{T_o} E \qquad (8.21)$$

Here T_s is the ocean temperature and T_o is the temperature of the outflow. Those temperatures and E may be easily estimated from observations of the tropics, and v as given by the equation is found to provide a good quantitative upper bound on hurricane wind speeds. Several factors, however, prevent most storms from achieving their maximum sustainable wind speed, or "potential intensity." Those include cooling of the sea surface by turbulent mixing that brings cold ocean water up to the surface and entropy consumption by dry air finding its way into the hurricane's core.

The thermodynamic cycle of a hurricane represents only a glimpse of the fascinating physics of hurricanes; more complete expositions are available in the resources given below. The transition of the tropical atmosphere from one with ordinary convective clouds and mixing-dominated entropy production to a system with powerful vortices and dissipation-driven entropy production remains a mysterious and inadequately studied phenomenon. This may be of more than academic interest, as increasing concentrations of greenhouse gases increase the thermodynamic disequilibrium of the tropical ocean–atmosphere system and thereby increase the intensity of hurricanes.

For more on hurricanes see K. Emanuel, "Divine Wind: The History and Science of Hurricanes", Oxford U. Press, New York (2005).

8.5 Athletes: the human engine

The fact that systems as complex as hurricanes can be understood in terms of the basic principles of thermodynamics leads one to wonder whether other "machines" capable of doing work can also be described in this manner. Here we want to examine the human engine; similar arguments of course apply to other living creatures. First a few facts about the energy consumption of humans. An adult at rest needs about 80 Watts to stay alive, or an energy input of 2 kWh per day (1 W = 1 J/s, 1 kWh = 3.6 megajoules). Performing mechanical work we need about $3 - 4$ kWh per day and 6 kWh per day for heavy work. Thus per year we need about 1500 kWh. The nutritional value of food items is 9 kWh/kg for butter, 2 kWh/kg for bread, and 1 kWh/kg for potatoes to list a few. In other words, we need a food supply of less than 1kg per day.

The efficiency of muscles to do work is about 20% for an average person and up to 40% for an athlete. If we were thermodynamic engines we would convert heat from a high temperature reservoir, our body, into work and transfer some of it to a low temperature reservoir. The latter is the ambient air around us and is on average about 20° C. Thus from $\varepsilon = 1 - T_c/T_h = 0.2$ we would need a body temperature of about 190° C. This is well above the boiling point of water, of which we are almost entirely composed. Clearly the work performed by our muscles cannot be done by way of a Carnot cycle. Rather, our body temperature is constant within a few degrees (fever and chill) $T_{body} = (37 \pm 3)°$ C suggesting that the chemical energy stored in our body is converted to muscular work isothermally.

Our bodies can store chemical energy in the form of mostly fat and some sugar. The amount of stored sugar is only enough for less than one hour of heavy work. More work is possible by burning fat which is more abundant in the body and has about twice the energy density of sugar. Our body consists of about 2×10^{13} cells which are supplied by the blood stream with nutrients, fat and sugar, and oxygen (attached to haemoglobin) to burn them and release energy. Because we know how much oxygen is needed to break down blood sugar and fat, a measurement of the oxygen consumption is an easy measure of the efficiency of the human engine.

Sugar burns very fast and produces a lot of wasted energy in the form of damaging free radicals. That is why our cells were designed to burn sugar only temporarily in times of great exigency, when the damage from free radicals is not as important as dealing with the emergency. Fat is the efficient form of energy that our cells should burn. A sugar rich diet builds up high levels of leptin which cause the hypothalamus (the energy controlling gland in the brain) to signal that the cells of their bodies should burn sugar, instead of fat.

Looking more closely at how our muscles perform work we can distinguish two mechanisms: the first is muscle contraction which produces a force which does work; this mechanism is similar to the discharge of a battery to do work. The isothermal conversion of chemical energy into work can reach an efficiency of up to 90%! The second mechanism is similar to charging the battery again. In the body this is done by burning fat or sugar in the presence of oxygen. This process produces a lot of heat and has a very small efficiency to do work.

Steady low level activity such as jogging or walking needs a chemical energy supply of about 1.5 kW with a concurrent oxygen supply of 4 liters per minute. Of this, about 300 Watts are available to do mechanical work. For short bursts of record breaking activity, like a 100 m sprint, the chemical energy reservoir can supply up to 100 kWs. If this activity leads to exhaustion the energy can be restored by a supply of 15 liters of oxygen in half an hour.

Remark 8.2. Here are a few facts about batteries to better understand the analogy to the human engine. In a battery, or electrochemical cell, energy is stored in the form of chemical bonds. When a cell is discharged, this energy is converted to an electrical current which in turn can do work. Efficiencies as high as 90% can be achieved in this type of energy conversion. A cell consists of two half-cells, each of which contain an electrode (an electrical conductor such as a piece of metal) and an electrolyte. The half-cells are connected by a salt-bridge - i.e. a conduit through which ions can pass freely. To understand the physics of this process, we consider the specific example of a Daniell cell where Zn and Cu are used as electrode materials immersed respectively, in an aqueous solution of $ZnSO_4$ and $CuSO_4$. When the two electrodes are connected to an external circuit, electrons are able to flow from the Zn (anode) to the Cu electrode (cathode). This flow of charge is due to differences in the electron affinity of Zn compared to Cu, i.e. their electron chemical potentials are not equal. Loss of electrons by the Zn electrode (oxidation) occurs according to the following half reaction: $Zn(s) - > Zn^{2+}(aq) + 2e^-$. As electrons flow out of this electrode, Zn ions dissolve into the solution. At the cathode, where electrons are deposited, Cu^{2+} ions, present in the solution, combine with this excess charge (reduction), forming $Cu(s)$; $Cu^{2+}(aq) + 2e- > Cu(s)$. As electrons flow into the cathode, Cu is extracted from the solution. Charge neutrality is maintained by transport of SO_4^{2-} and Zn^{2+} ions across the salt bridge. As an aside, the working mechanism of a mammalian cell also involves charged ions that are crucial for the transmission of signals.

8.6   Thermodynamics in economics

Like physical systems, economic systems are large, multifaceted, and show large-scale trends that go beyond the detailed activities of individual players (the equivalent of atoms and molecules) and their mutual activities and interactions. This suggests that a macroscopic description in terms of a few variables should also be possible and useful. To do this we need to identify a suitable set of variables that describe economic systems in general and on a long time scale. The latter restriction is akin to what we always do in thermodynamics, namely that such a description is restricted to systems in equilibrium and to processes close to equilibrium, i.e. quasi-static processes. One might think that this is not achievable because economic systems are always in flux and fluctuate widely, such as in economic boom and bust times. However, the overall growth of a national economy is typically at a rate of a few percent per year so is the economy of most individuals who have a fixed income, adjusted for inflation and performance, which again varies at the percent level from year to year. It is therefore suggestive that one should first of all study the economics of a system in equilibrium, hence the use of thermodynamics.

As we did with physical systems we postulate that there is a small number of macroscopic, extensive variables for which a fundamental relation exists for each economic system.

Economic theory introduces a utility function to choose to buy one good over another in an analogous way to the internal energy U determining which processes can proceed. Thus we identify U as the utility function, measured in some monetary values. Another quantity in economics is wealth which has two contributions

$$W = \boldsymbol{\lambda} \cdot \mathbf{M} + \mathbf{p} \cdot \mathbf{N} \tag{8.22}$$

Here $\boldsymbol{\lambda}$ and M are vectors containing the value, i.e. the price one would get relative to some standard (like the gold standard) and the amount of money held in different currencies. Likewise, $\mathbf{p}$ and $\mathbf{N}$ represent prices and numbers of goods respectively. The difference between the utility function and wealth is called the surplus

$$\Psi = U - W \tag{8.23}$$

or

$$U = \Psi + W \tag{8.24}$$
$$= \Psi + \boldsymbol{\lambda} \cdot \mathbf{M} + \mathbf{p} \cdot \mathbf{N} \tag{8.25}$$

We can quantify the term "poverty" as meaning zero surplus and zero economic activity. Together with the fact that a surplus by definition is positive, we identify the surplus with TS and get

$$U = U(S, M, N) \tag{8.26}$$
$$= T_e S + \boldsymbol{\lambda} \cdot \mathbf{M} + \mathbf{p} \cdot \mathbf{N} \tag{8.27}$$

This is our fundamental relation for a simple economic system. What is the meaning of S in economics? Instead of using the term disorder we call it a measure of available leisure in the sense that with zero surplus there is no room for leisure. We also measure leisure in monetary units so that the intensive variable T_e is dimensionless.

The fundamental relation in differential form reads

$$dU = T_e dS + \lambda \cdot dM + p \cdot dN \tag{8.28}$$

so that

$$T_e = \frac{\partial U}{\partial S}\Big|_{M,N} \tag{8.29}$$

$$\lambda_i = \frac{\partial U}{\partial N_i}\Big|_{S,M,N_j} \tag{8.30}$$

$$p_i = \frac{\partial U}{\partial S}\Big|_{S,M,N_j} \tag{8.31}$$

are the intensive variables. Of particular interest is the meaning of the economic temperature T_e: it is a measure of the increase in utility with increasing leisure in the system keeping the amount of money and the number of goods in the system constant.

Let us look at a situation where S and M are fixed and two commodities are exchanged under the condition that this exchange will minimize the utility function. We get

$$dU = p_1 dN_1 + p_2 dN_2 = 0 \tag{8.32}$$

or

$$\frac{1}{p_i}\frac{\partial U}{\partial N_i}\Big|_{S,M,N_j} = const. \tag{8.33}$$

This recovers a profound statement from 'marginalist' economic theory: a consumer will purchase several goods subject to the condition that the ratio of any good in use to its price p takes on a common value.

Let us next couple a small economy to a larger one and find out into which equilibrium state the two systems will settle. According to Chapter 4 all intensive variables will be equal in the two systems. Because the larger economy will adjust very little if exposed to a smaller economy, the latter will adjust its T_e, λ_i, and p_i to the larger system. Only trade barriers, i.e. 'walls' that are impenetrable to the flow of money or goods can prevent that!

In Section 1.2.1 we introduced the compressibility and expansion coefficients

$$\kappa_T = -\frac{1}{V}\frac{\partial V}{\partial P}\Big|_T \tag{8.34}$$

$$\alpha = \frac{1}{V}\frac{\partial V}{\partial T}\Big|_P \tag{8.35}$$

as useful and easily measurable quantities for the characterization of systems: κ_T is a measure of the relative decrease of volume as pressure is increased. By its very definition it must be positive for a system in equilibrium; if it were negative the system would implode. Likewise, α is a measure of the relative volume increase with temperature. Whereas for most systems it is positive, this is not always the case. As an example, lowering the temperature of water to the point of freezing the volume decreases to about 4° C and then increases again.

We now adopt the definitions (8.34) to economic systems

$$\kappa_T^{(e)} = \frac{1}{M}\frac{\partial M}{\partial \lambda}\Big|_{T_e}$$

$$\alpha^{(e)} = \frac{1}{M}\frac{\partial M}{\partial T_e}\Big|_{\lambda} \tag{8.36}$$

Note that in the definition of $\kappa_T^{(e)}$ we do not have a minus sign as in κ_T because in the fundamental relation of physical systems the work term is $-PdV$ whereas in an economic system its is λdM. $\kappa_T^{(e)}$ is a measure for the relative change in the amount of money in the economy with a change of price or value for that currency. A situation where the amount of money increases as its price decreases is called inflation, $\kappa_T^{(e)}$ is negative, and if it is not checked by other means such as raising interest rates, we have run-away inflation: the system is unstable. Thus for an economy to be stable $\kappa_T^{(e)}$ must be positive just as for the compressibility of physical systems. What about $\alpha^{(e)}$? It would be positive, i.e. the amount of money increases as the index of leisure increases, if you hired an investment banker to make money for you without you having to work for it: the richer you get the more leisure time you have. $\alpha^{(e)}$ would also be positive in inflationary times when you might make less money despite the fact that you work more. But $\alpha^{(e)}$ can also be negative, namely in a situation where you live beyond your means, i.e. you take more leisure time at the expense of diminishing your savings. Again, the situation is similar to physical systems.

To go further with the thermodynamics one needs empirical equations of state such as the ideal gas law $PV = NRT$. Its analogy in economics would be

$$\lambda M = N c_e T^{(e)} \tag{8.37}$$

with c_e some constant. From (8.37) we get

$$\kappa_T^{(e)} = \frac{1}{\lambda}$$

$$\alpha^{(e)} = \frac{1}{T_e} \tag{8.38}$$

Both make sense for a stable economy: the relative change in the amount of money with raising value of the currency decreases as the currency gets pricier; and similarly for $\alpha^{(e)}$.

It seems that the task of measuring equations of state or quantities like $\kappa_T^{(e)}$ and $\alpha^{(e)}$ has not been taken up in economic theory. Perhaps the next generation of thermodynamicists will rise to the challenge. For more on the thermodynamics of economics, see the paper by Wayne M. Saslow on "An economic analogy to thermodynamics" in *American Journal of Physics 71 (1999) pp. 1239-1247.*

8.7 Problems

Problem 8.1. *Efficiency of the Brayton or Joule cycle*

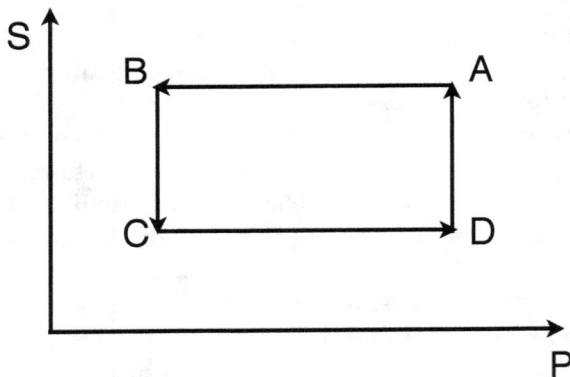

Fig. 8.10   The Brayton engine.

(a) *Assuming that the working fluid is an ideal gas calculate heat transfers and work for each of the four processes.*
(b) *Calculate the engine efficiency.*
(c) *What is the fraction of heat wasted. Where does it go?*

Problem 8.2. *Efficiency of the Stirling cycle*

(a) *Assuming that the working fluid is an ideal gas calculate heat transfers and work for each of the four processes.*
(b) *Calculate the engine efficiency.*

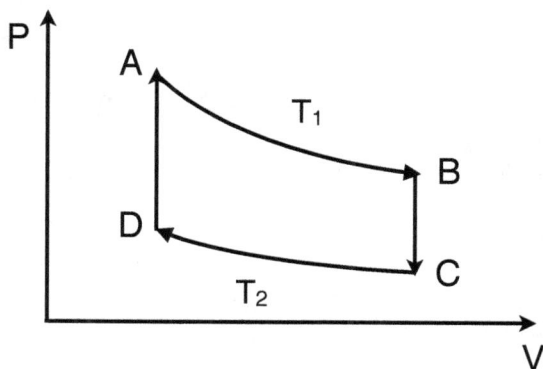

Fig. 8.11   The Stirling engine.

(c) What is the fraction of heat wasted. Where does it go?

(d) How does this compare to the Carnot cycle?

Problem 8.3. *Efficiency of the Ericsson cycle, a simple model for a regenerative gas turbine. In common with the Carnot and Stirling cycles, the heat exchange with the temperature reservoirs occurs in isothermal expansion and compression.*

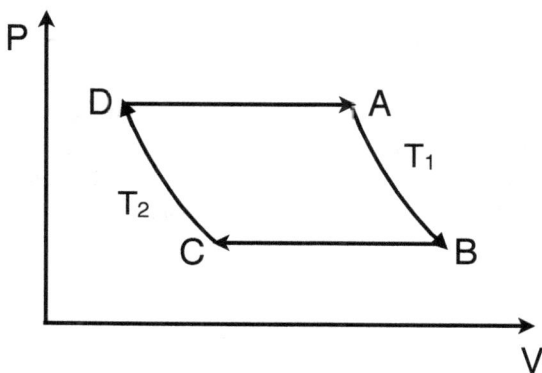

Fig. 8.12   The Ericsson engine.

(a) Assuming that the working fluid is an ideal gas, calculate heat transfers

and work for each of the four processes.

(b) Show that work and heat transfer in the two isobaric processes are equal but of opposite signs.

(c) Show that $V_b/V_a = V_c/V_d$ and $Q_{a \to b}/ Q_{a \to b} = T_1/T_2$

(d) Calculate the engine efficiency.

(e) What is the fraction of heat wasted. Where does it go?

Problem 8.4.

Two identical systems with constant specific heat are initially at temperatures T_1 and T_2. They are used as the source and sink of a Carnot cycle and no other heat source is available.

(a) Show that the final temperature is $T = \sqrt{T_1 T_2}$ when all possible work is extracted.

(b) What is the maximum work obtainable?

Problem 8.5. *A person is wrapped in an adiabatic suit so that no heat generated by metabolism can escape. Assume that the rate at which heat is produced by metabolism is* 10 *kJ/kg/hr. Also assume that in the range of the body temperature the heat capacity is constant* $C = 4.18$ *J/g/K. What is the body temperature after three hours?*

Chapter Summary

(1) The Carnot cycle is introduced. Both the ideal and realistic cases are considered.

(2) Arguments are given for the maximum work theorem. This leads to a definition of thermodynamic efficiency.

(3) The thermodynamics of several systems, including engines, refrigerators, heat pumps, and cyclones are discussed.

Chapter 9

Stability of Thermodynamic Systems

9.1 Macroscopic motion

The fact that macroscopic systems are comprised of molecules and atoms (which move incessantly) is of no direct relevance to thermodynamics— microscopic degrees of freedom are fully accounted for through entropy and heat transfer. What is not clear at this point is what effect, if any, macroscopic motion has on the thermodynamics of a given system (or subsystem). Here we wish to examine what kind of motion macroscopic subsystems of a given system can undergo while remaining in thermal equilibrium.

We divide the system into many small, but still macroscopic, subsystems. We assign a mass M_a, energy E_a, and momentum $\mathbf{P}_a$ to each. The entropy of each subsystem is a function of its internal energy (which is the difference between its total energy and its macroscopic kinetic energy), so that the entropy of the whole system is given by

$$S = \sum_a S_a(E_a - \frac{P_a^2}{2M_a}, V_a, N_a) \tag{9.1}$$

The fact that the kinetic energy is irrelevant is a consequence of Galilean invariance: internal motion in a system is the same in all inertial reference frames. Recall that an inertial reference frame is one that is *not* undergoing any form of acceleration.

We assume that the system as a whole is isolated so that in addition to its total energy, its linear momentum and angular momentum are conserved

$$\sum_a E_a = const.$$

$$\sum_a \mathbf{P}_a = const.$$

$$\sum_a \mathbf{r}_a \times \mathbf{P}_a = const. \tag{9.2}$$

In equilibrium the entropy must be a maximum as a function of the momenta, provided the auxiliary conditions above are fulfilled. Using the method of Lagrange multipliers we set the derivatives of

$$\sum_a [S_a + \mathbf{a} \cdot \mathbf{P}_a + \mathbf{b} \cdot (\mathbf{r}_a \times \mathbf{P}_a)] \tag{9.3}$$

with respect to $\mathbf{P}_a$ to zero. For the entropy we find

$$\frac{\partial}{\partial \mathbf{P}_a} S_a (E_a - \frac{P_a^2}{2M_a}) = -\frac{\mathbf{v}_a}{T}$$

$$\mathbf{v}_a = \frac{\mathbf{P}_a}{M_a} \tag{9.4}$$

Thus differentiation of (9.3) yields

$$-\frac{\mathbf{v}_a}{T} + \mathbf{a} + \mathbf{b} \times \mathbf{P}_a = 0 \tag{9.5}$$

or, with constant vectors $\mathbf{u} = T\mathbf{a}$ and $\Omega = T\mathbf{b}$

$$\mathbf{v}_a = \mathbf{u} + \Omega \times \mathbf{r}_a \tag{9.6}$$

This result has an intuitive interpretation: if the velocities of all macroscopic parts are given by this equation, then the system as a whole is subject to uniform linear motion and rotation. In other words, in thermal equilibrium an isolated system can only undergo such motion, and internal macroscopic motion of its parts is impossible! Thus we can deal with thermodynamics by assuming the system is at rest.

This insight leads us to a simple proof that the temperature as defined by the Kelvin scale is positive. We again consider an isolated system at rest. If its temperature were negative we would need the entropy to increase for decreasing internal energy. As a result the system would spontaneously disintegrate into many subsystems so that the argument of the entropy S_a for each system gets as small as possible. In other words, if $T < 0$ were possible we could not have any system in thermal equilibrium.

9.2 Thermodynamic inequalities

So far we have established and repeatedly used several extremum principles to define and find the equilibrium states of a thermodynamic system. We started by arguing that in equilibrium the entropy $S(U, V, N)$ must attain a maximum, implying that its first differential must be zero, $dS = 0$. Likewise, we found that all free energies or thermodynamic functions U, F, H, G must be extrema, so that their first differentials are also zero in equilibrium. These conditions only establish the fact that the equilibrium states are extrema. To confirm whether the extremum for the entropy or the internal energy is a maximum or a minimum, respectively, we need to establish that the second derivatives are either positive or negative, i.e. $d^2S < 0$, $d^2U > 0$, etc. These conditions will impose inequalities on the second derivatives, which we intend to derive now. Let us therefore expand the internal energy (for a system with constant mole number) in a multivariate Taylor expansion

$$U(S + dS, V + dV, N + dN) = U(S, V, N) + dU + d^2U \qquad (9.7)$$

$$dU = \frac{\partial U}{\partial S}|_{V,N} dS + \frac{\partial U}{\partial V}|_{S,N} dV = TdS - PdV$$

$$d^2U = \frac{1}{2}\left[\frac{\partial^2 U}{\partial S^2}|_{V,N}(dS)^2 + 2\frac{\partial^2 U}{\partial S \partial V}|_N dS dV + \frac{\partial^2 U}{\partial V^2}|_{S,N}(dV)^2\right] \qquad (9.8)$$

For the quadratic form d^2U to be positive definite its coefficients must satisfy the conditions

$$\frac{\partial^2 U}{\partial S^2}|_{V,N} > 0 \qquad (9.9)$$

$$\frac{\partial^2 U}{\partial S^2}|_{V,N}\frac{\partial^2 U}{\partial V^2}|_{S,N} - \left[\frac{\partial^2 U}{\partial S \partial V}|_N\right]^2 > 0 \qquad (9.10)$$

The first condition can be written

$$\frac{\partial^2 U}{\partial S^2}|_{V,N} = \left[\frac{\partial T}{\partial S}|_{V,N}\right] = \frac{T}{Nc_V} > 0 \qquad (9.11)$$

implying that for a thermodynamic equilibrium state to be stable its specific heat at constant volume must be positive

$$c_V > 0 \qquad (9.12)$$

Transfer of heat into a system at constant volume and mole number must lead to an increase in temperature. If this were not the case the system would cool down to absolute zero upon the addition of heat.

In the second inequality we introduce intensive variables and get

$$\frac{\partial T}{\partial S}|_{V,N}\left[-\frac{\partial P}{\partial V}|_{S,N}\right] - \left[-\frac{\partial P}{\partial S}|_{V,N}\right]^2 > 0 \tag{9.13}$$

Using Maxwell relations and other tricks from Chapter 7, we can transform this expression into

$$\frac{T}{Nc_V}\frac{\partial P}{\partial V}|_{T,N} = -\frac{T}{NV\kappa_T} < 0 \tag{9.14}$$

implying that in stable equilibrium the isothermal compressibility must be positive

$$\kappa_T > 0 \tag{9.15}$$

Reducing the volume of a system in equilibrium at a fixed temperature must result in an increase of pressure. If that were not the case the system would implode. Equations (9.12) and (9.14) are called thermodynamic inequalities; they are the criteria that guarantee thermal and mechanical stability of equilibrium states. With (7.26) we also have

$$c_P = c_V + T\frac{V}{N}\frac{\alpha^2}{\kappa_T} > c_V > 0 \tag{9.16}$$

The reason why the specific heat at constant pressure is always larger than at constant volume is that at constant pressure some of the heat transferred to the system must be used to do work to change the volume. Similarly (7.35) implies

$$\kappa_s = \frac{c_V}{c_P}\kappa_T < \kappa_T \tag{9.17}$$

The fact that c_V and c_P are positive implies that the internal energy increases monotonically as a function of T at constant volume and the enthalpy does the same at constant pressure. On the other hand, the entropy grows monotonically with temperature both at constant volume and at constant pressure.

Remark 9.1. We have derived the thermodynamic inequalities for homogeneous systems for which these criteria are true both for the system as a whole and locally for any small but still macroscopic part of it. For inhomogeneous systems this need not be true globally. An example is a system such as a planet that is held together by gravitational forces which becomes denser towards the center. In such systems the specific heat of the system as a whole may become negative, i.e. the system gets hotter as its energy decreases. The reason is that in a gravitational body the decrease in energy can be due to more gravitational potential energy which is negative. Locally, however, the thermodynamic inequalities still hold!

Remark 9.2. We have derived the stability criteria for small deviations from the equilibrium state so that an expansion to second order is enough. Let us now consider a situation where the Gibbs free energy (or any other thermodynamic potential) has a local minimum, say as a function of temperature or pressure, so that for small deviations the system returns to the minimum. However, if the deviations grow large the system will jump into a deeper minimum that, for simplicity, we assume to be the global minimum of the Gibbs free energy. In this case, the local minimum was NOT the stable equilibrium but a metastable state. How long a system will remain in a metastable state is a function of the barrier height over which the system has to go to reach the stable state. This transition is usually triggered by a fluctuation in a small, but still macroscopic region of the system and then propagates throughout the system to bring it into stable equilibrium. This is what happens to diamond vs. graphite: diamond is metastable and graphite is the stable equilibrium form of carbon at room temperature and atmospheric pressure. We further discuss these types of phase transitions in Chapter 10.

9.3 Fluctuations and the principle of Le Chatelier and Braun

Before we formulate the principle of Le Chatelier and Braun we look at a simple example: consider a gas in a diathermal cylinder fitted with a moveable piston. This system is immersed in a heat and pressure reservoir. Imagine that due to a fluctuation in the gas the piston moves slightly outward, increasing the volume and decreasing the pressure in the gas below that of the pressure reservoir. The latter will therefore establish a pressure gradient across the piston forcing it to move back to its equilibrium position. This is the principle of Le Chatelier: any fluctuations around a stable equilibrium in a system will cause a reaction that will drive the system back to equilibrium. So far we have assumed that the heat reservoir reacts so fast that the temperature is keep constant during the instantaneous expansion. If this is not the case, the small decrease in pressure will trigger a small cooling of the gas. Consequently there will be a small heat flow from the heat reservoir to re-establish the original temperature. But this heat flow results in a pressure increase

$$dP = \frac{1}{T}\frac{\partial P}{\partial S}|_V dQ = \frac{\alpha}{NT^2 c_V \kappa_T} dQ \qquad (9.18)$$

Result: the secondary effect of the initial pressure fluctuation, i.e. the lowering of the temperature, will also cause a response to counteract the initial pressure change driving the system back to equilibrium. This is the principle of Le Chatelier and Braun.

We formulate both principles using the internal energy of a simple system, $U = U(S, V, N)$, in contact with a temperature and pressure reservoir. Assume that the entropy of the total system, system plus reservoirs, undergoes a fluctuation dS^f resulting in a change in the temperature of the system

$$dT^f = \frac{\partial T}{\partial S} dS^f \tag{9.19}$$

But this entropy fluctuation also causes a fluctuation in the pressure

$$dP^f = \frac{\partial P}{\partial S} dS^f \tag{9.20}$$

To get the response of the reservoirs, dS^r and dV^r, we minimize the total internal energy at constant total entropy, and get

$$d(U + U^r) = dT^f dS^r + dP^f dV^r \le 0 \tag{9.21}$$

But since dS^r and dV^r are independent, we have

$$dT^f dS^r \le 0$$
$$dP^f dV^r \le 0 \tag{9.22}$$

and using (9.19) and (9.20)

$$\frac{\partial T}{\partial S} dS^f dS^r \le 0$$
$$\frac{\partial P}{\partial S} dS^f dV^r \le 0 \tag{9.23}$$

Multiplying the first inequality by $\partial T/\partial S$ we obtain

$$\frac{\partial T}{\partial S} \frac{\partial T}{\partial S} dS^f dS^r \le 0 \tag{9.24}$$

or

$$dT^f dT^{r(1)} \le 0 \tag{9.25}$$

Thus the response dS^r of the reservoir produces a change in the temperature of the reservoir that is of opposite sign to the initial change in the system's temperature. This is Le Chatelier's principle.

Turning now to the second inequality in (9.23) we use the Maxwell relation $\partial P/\partial S = \partial T/\partial V$ and multiply with the positive quantity $\partial T/\partial S$ to get

$$\frac{\partial T}{\partial S}dS^f \frac{\partial T}{\partial V}dV^r \leq 0 \tag{9.26}$$

or

$$dT^f dT^{r(2)} \leq 0 \tag{9.27}$$

Thus the response dV^r of the reservoir produces a change in the temperature of the reservoir that is of opposite sign to the initial change in the system's temperature. This is the principle of Le Chatelier and Braun.

In summary: any fluctuation in a system will produce responses in both the system itself and any reservoirs to which it is coupled so as to restore the system to stable equilibrium.

Chapter Summary

(1) Stability of a system in equilibrium demands that $c_P > c_V > 0$ and $\kappa_T > \kappa_s > 0$.
(2) Principle of Le Chatelier: any fluctuations around a stable equilibrium in a system will cause a reaction that will drive the system back to equilibrium.
(3) Principle of Le Chatelier and Braun: any fluctuations in a system will produce responses in the system and in reservoirs to which it is coupled that restore the system to stable equilibrium.

Chapter 10

Phase Transitions

Our discussion to this point has been focused primarily on homogeneous systems. By homogeneous we mean a single chemical component in a single phase, e.g. gas, liquid, or solid. The exception to this was our discussion of water droplets in Section 6.4.1. We now turn our attention to inhomogeneous systems. Inhomogeneous systems are common in everyday life. A glass of ice-water is a perfect example of a single species existing simultaneously in several different phases: liquid water, solid ice, and water vapor. What can thermodynamics tell us about these types of systems? In this example, the answer is actually "very little". Thermodynamics is concerned with systems that are at or near equilibrium. The fact that the ice cubes in the water will likely melt, and the subsequent liquid will evaporate if left alone is a strong indication that this system is not at equilibrium. Given this information, we can ask the question of whether or not it is possible to have an *equilibrium* multiphase system? That is to say if left undisturbed, will a system with more than one phase present remain in that state indefinitely? As we shall see, the answer to this question is yes, provided the thermodynamic conditions are right.

To frame our discussion, we conduct an experiment where we place water vapor at a fraction of an atmosphere into a vessel with a movable piston. Let us now reduce the volume by moving the piston inward, increasing the pressure of the vapor. As we continue decreasing the volume, some of the vapor will condense into liquid water at the bottom of the vessel. The system is now inhomogeneous, consisting of a low density vapor phase and a high density liquid phase. This can be viewed as a better use of volume. By condensing some of the vapor into a dense liquid the remaining vapor has a bigger volume at its disposal. Eventually all the vapor has condensed and the piston is in contact with the liquid resulting again in a homogeneous system. The relationship between how much pressure must be applied at what temperature is governed by the equation of state of the system.

The equilibrium state of a homogeneous system is determined by specifying its volume, mole number, and internal energy, and by the entropy as a function of these variables. However, as we just saw, there is no guarantee that for any set of these variables the system is homogeneous. Two states of a material that are homogeneous at different densities, and can co-exist, are called phases. The separation of a homogeneous state into two phases is called a phase transition.

Because the phases of a material are in thermal and mechanical contact with each other through their interface, their temperatures and pressures must be equal

$$T_1 = T_2 = T$$
$$P_1 = P_2 = P \tag{10.1}$$

Likewise, the chemical potentials must be equal

$$\mu_1(T, P) = \mu_2(T, P) \tag{10.2}$$

This condition implies that T and P are not independent,

$$P = P(T) \tag{10.3}$$

and the two phases can only co-exist along a line in the $P - T$ plane; it is called the co-existence line (see the left panel in Figure 10.1).

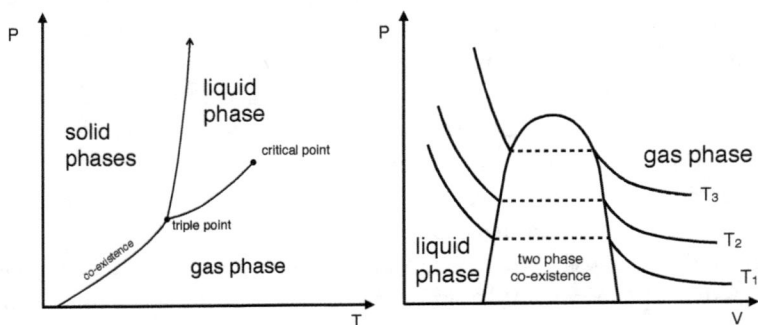

Fig. 10.1   Schematic of a phase diagram (left) in the $P - T$ plane showing three phases. On the right we show isotherms in the $P - V$ plane together with the co-existence region of the gaseous and liquid phases.

For any point in the $P - T$ plane off the co-existence line, the system is homogeneous in one of the two phases. If one performs an experiment in which, for instance, the pressure is held constant and the temperature is lowered, phase separation starts when the co-existence line is reached. The system is in the other phase to the left of the co-existence line. To understand this we note that the molar volumes (or densities) are vastly different in the gas and in the liquid. In other words, as we reach the co-existence line we must start to reduce the volume and will see a continuous transformation of the dilute into the dense phase. On the other hand, a line in the $P - T$ plane corresponds to an area of finite extent (for fixed mole number) in the $T - V$ plane (see Figure 10.2).

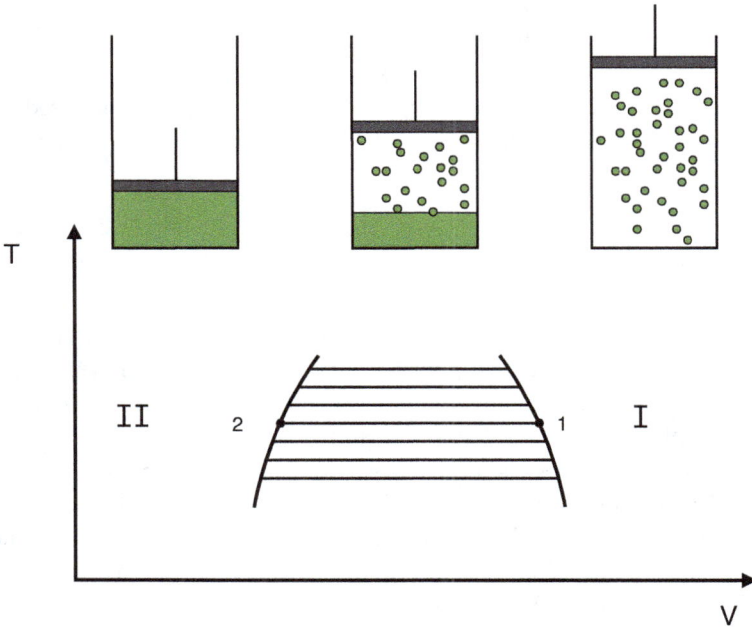

Fig. 10.2   Co-existence region in the $T-V$ plane together with an illustration of reducing the volume.

Points in area I are homogeneous and are the dilute phase. We reduce the volume at constant temperature by pushing a piston, thus reducing the volume at constant pressure. Phase separation starts at "1" and throughout the shaded region the liquid phase of higher density grows at constant pressure because $P = P(T)$ on the co-existence curve. At "2" the phase transition is complete - in area II we again have a single homogeneous phase, but this time it is the condensed liquid phase. Moving into the condensed region, we must increase the pressure to compress the liquid. Corresponding isotherms in the $P - V$ plane are plotted in the right panel of Figure 10.1.

How much condensed vs. dilute phase do we have at a given volume in the co-existence region "1" to "2"? Because the amount of material is conserved in the phase transition, we can define the mole fraction of dilute and condensed phase as

$$x_g = \frac{N_g}{N}$$

$$x_l = \frac{N_l}{N} \tag{10.4}$$

with $x_g + x_l = 1$. The total volume along the path from "1" to "2" is

$$V = Nv = Nx_g v_g + Nx_l v_l \tag{10.5}$$

where v_g and v_l are the molar volumes of the dilute and condensed phase, respectively, i.e. the volumes at points "1" and "2". Thus we have

$$x_g = \frac{v - v_l}{v_g - v_l}$$

$$x_l = \frac{v_l - v}{v_g - v_l} \tag{10.6}$$

This is called the lever rule: along an isotherm inside the co-existence region, the mole fractions of the two phases in a volume V are given by its fractional distance from the opposite ends of the co-existence region.

What we discussed so far about the co-existence of the gas phase with the liquid phase is of course true for any two phases with a discontinuous phase transition. Thus there is a co-existence line for the gas phase with the solid phase and for the liquid and the solid phase (see Figure 10.1).

Unlike the solid-liquid co-existence line, the boundary between the gas and liquid phase is of finite extent. The termination point at the upper end, for the highest temperature and highest pressure of co-existence, is called the critical point; new physics happens there which we will discuss in Section 10.5. The termination point at the lower end, for the lowest temperature and lowest pressure, is called the triple point. Here three phases - gas, liquid, and solid - co-exist (this is analogous to two-phase co-existence). The fact that the three phases are in equilibrium with each other implies that their temperatures, pressures, and chemical potentials must be equal

$$T_1 = T_2 = T_3 = T$$

$$P_1 = P_2 = P_3 = P$$

$$\mu_1(T, P) = \mu_2(T, P) = \mu_3(T, P) \tag{10.7}$$

The additional condition that the third phase must also be in equilibrium with the other two phases implies that three phases can only be in co-existence at a single point, i.e. for one temperature T_{triple}, one pressure P_{triple}, and one molar volume v_{triple} (or one density v_{triple}^{-1}). For any substance or material the triple point is unique. This uniqueness is behind the choice of a specific triple point, namely that of water, for the calibration of the Kelvin temperature scale, as we discussed in Chapter 1 in connection with the constant volume gas thermometer. For further elucidation we show in Figure 10.3 the $P-T$ and $T-V$ planes for three phases, indicating the various co-existence regions. For $P < P_{\text{triple}}$, the solid will change directly to the gas.

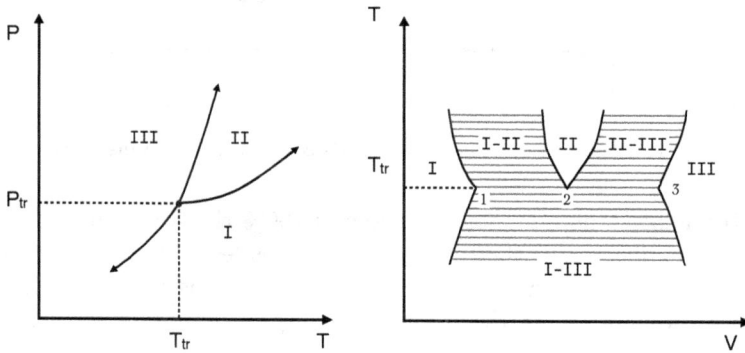

Fig. 10.3   Phase diagram of three phases in the $P-T$ and $T-V$ planes.

In Figure 10.4 we again look at typical phase diagram to introduce a few terms relevant to phase transitions, namely the pairs condensation-evaporation, sublimation-deposition, freezing-melting.

Remark 10.1. As the phase diagram of water shows (Figure 10.5), water vapor and liquid water at atmospheric pressure only co-exist at 100° C for a range of volumes along the co-existence line. At first sight this seems to contradict everyday experience in a profound way. We all know that at room temperature you can have a glass of water with some water vapor above the water surface, i.e. there seems to be co-existence over a whole range of pressures and temperatures where there should not be one. What is the resolution of this contradiction? Well, what counts is not the absolute pressure of air above the surface of the water but the partial pressure of the water vapor which at room temperature is only a fraction of an atmosphere.

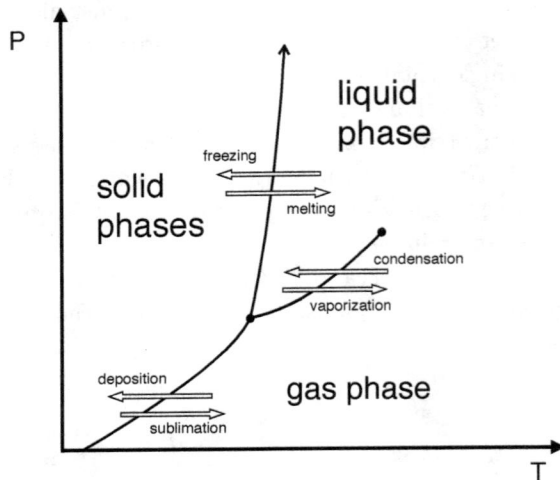

Fig. 10.4   Illustration of pairs of words connected with transitions.

So, if the relative humidity in the room is 100% the water in the glass is in stable equilibrium. If the relative humidity is less than that, water will evaporate to increase the vapor pressure. But as this has to be done for the whole volume of the room (and beyond) the water in the glass will eventually evaporate completely.

10.1   Latent heat

For a fluid system, the transition from one phase to another is accompanied by the transfer of heat, as evidenced in every day life when boiling water. In this case heat must be transferred into the liquid to break up the bonds between the water molecules to set them free. This heat is called latent heat (per mole), q. It is positive when heat must be transferred into the system (evaporation or boiling) and is negative when heat has to be extracted (condensation). Now, the phase transition happens at a given pressure and temperature. But the heat transfer in a process at constant pressure is equal to the change in enthalpy, as we have seen in Section 6.3. Thus we have for the transition from phase "1" to phase "2"

$$q = h_2 - h_1 \tag{10.8}$$

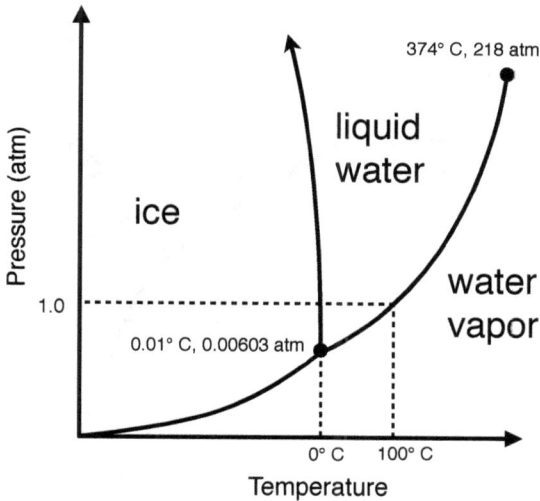

Fig. 10.5 Low pressure region of the phase diagram of water.

On the other hand, the chemical potential is the Gibbs free energy per mole because $G = U - TS + PV = H - TS = N\mu$. Thus the equality of the chemical potentials in the two phases, $\mu_1 = \mu_2$, implies

$$h_1 - Ts_1 = h_2 - Ts_2 \tag{10.9}$$

so that the latent heat becomes

$$q = T(s_2 - s_1) \tag{10.10}$$

If the second phase is vapor and the first phase is a liquid then the entropy of the vapor is of course much larger than that of the liquid (more disorder) and latent heat is positive and must be transferred to achieve evaporation.

In Figure 10.6 we show the chemical potentials of three phases as a function of temperature at constant pressure. For temperatures below T_{melt} the solid is stable, as its chemical potential, or molar Gibbs free energy, is lower because the Gibbs free energy must be a minimum at equilibrium. At T_{melt}, the chemical potentials of the solid phase and the liquid phase cross and the solid is in co-existence with the the second phase (the liquid). Between T_{melt} and T_{boil} the liquid phase is stable. Finally, at T_{boil}, again two phases co-exist, this time the liquid and the gas, and above T_{boil} only the gas is stable. Also note that at the crossing points we have for the

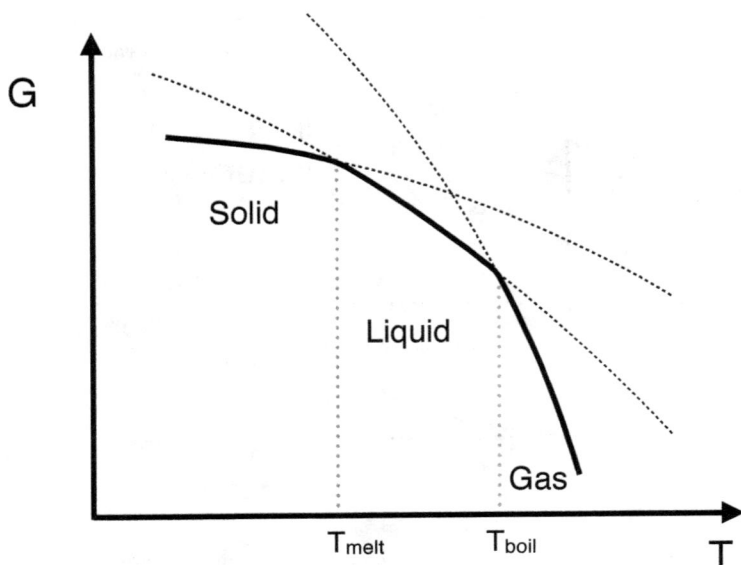

Fig. 10.6 The Gibbs free energy for three phases as a function of temperature. The three segments that minimize the Gibbs free energy are thermodynamically stable.

slopes of the two chemical potentials, i.e. for the molar entropies of the gas and liquid phases,

$$s_{gas} = -\frac{\partial \mu_{gas}}{\partial T}\Big|_P > s_{liquid} = -\frac{\partial \mu_{liquid}}{\partial T}\Big|_P \qquad (10.11)$$

implying, again, that the latent heat $q = T(s_{gas} - s_{liquid})$ is positive: when for increasing temperature a system undergoes a phase transition (from the liquid into the gas) it will absorb heat (the heat of vaporization), as required by Le Chatelier's principle!

A phase transition in which a thermodynamic free energy is non-analytic as a function of one of its variables, so that its first derivatives have a discontinuity (such as the entropy and the molar volume as derivatives of the Gibbs free energy) are called discontinuous or first order phase transitions. Such phase transitions are accompanied by a latent heat. Continuous or second (and higher) order phase transitions are characterized by continuous first derivatives and by a discontinuous second or higher derivative. As an example of a second order phase transition, at the paramagnetic to ferromagnetic phase transition of iron, the magnetization as a derivative of the free energy with respect to the magnetic field is continuous but the second derivative, the magnetic susceptibility, is discontinuous. There is no latent heat in continuous transitions and two phase co-existence is not possible.

10.2   Clausius-Clapeyron equation

As we shall see, in addition to the location of phase transitions, the slope of a transition (with respect to another thermodynamic variable) can have important implications for equilibrium structures. We will now calculate this slope. We again start with the phase co-existence condition (10.2) and take the derivative on both sides with respect to T, remembering that along the co-existence line $P = P(T)$

$$\frac{\partial \mu_1}{\partial T} + \frac{\partial \mu_1}{\partial P}\frac{dP}{dT} = \frac{\partial \mu_2}{\partial T} + \frac{\partial \mu_2}{\partial P}\frac{dP}{dT} \tag{10.12}$$

Because $\partial \mu/\partial T|_P = -s$ and $\partial \mu/\partial P|_T = v$ we get

$$\frac{dP}{dT} = \frac{s_1 - s_2}{v_1 - v_2} \tag{10.13}$$

in terms of the molar entropies and volumes of the two phases, or, in terms of the latent heat (10.10)

$$\frac{dP}{dT} = \frac{q}{T(v_2 - v_1)} \tag{10.14}$$

This is the Clausius-Clapeyron equation: it determines the change of pressure as a function of temperature along the co-existence curve. Inverting this equation

$$\frac{dT}{dP} = \frac{T(v_2 - v_1)}{q} \tag{10.15}$$

We see that with increasing pressure a liquid (phase "1" with molar volume v_1) increases because the molar volume of the gas phase is always much larger that that of the liquid, $v_2 \gg v_1$. This is again in agreement with Le Chatelier's principle because as the pressure increases more gas particles transfer to the liquid thus reducing the pressure, or in Le Chatelier's language, the system counteracts the externally applied constraint (higher pressure). This enhancement of the boiling point with pressure is the phenomenon underlying the pressure cooking invented by Count Rumford.

Applying the inverted Clausius-Clapeyron equation to freezing we see that the freezing temperature will be increased or decreased depending on whether the liquid volume v_1 is smaller or larger than the solid volume v_2. For most materials $v_1 > v_2$, with the notable exception of water where

$v_2 > v_1$. That is why ice floats on water and water always freezes from the top down.

Remark 10.2. The fact that the molar volume of ice is larger than that of liquid water has far-reaching consequences. If it were the other way around, lakes and rivers would freeze from the bottom up. The chance that a deep lake would ever thaw again in Spring would be non-existent. As a lake freezes from the surface down, ice builds up a thermal shield against colder air temperatures providing a safe heaven for aquatic life.

On the other hand, a bucket or barrel of water freezing from the top compresses the water below to much higher pressures that will eventually burst the bottom. This can, by the way, be easily prevented by putting a good-sized wooden pole into the barrel, as it will be compressed more easily making room for the expanding ice.

Returning to the co-existence of a gas with either its liquid or solid, we note that the molar volume of the gas is always much larger than that of the solid or liquid, $v_2 \gg v_1$, so that we can neglect it in the Clausius-Clapeyron equation. Treating the gas as ideal we have $v_2 = RT/P$ and get from (10.14)

$$\frac{dP}{dT} = \frac{qP}{RT^2} \tag{10.16}$$

$$\frac{d\ln P}{dT} = \frac{q}{RT^2}$$

This is sometimes called the Clapeyron equation. Over temperature intervals along the co-existence curve where the latent heat is more or less constant we get

$$P = P_0 \exp(-q/RT) \tag{10.17}$$

Thus the pressure decreases along the co-existence curve for increasing temperatures.

Example 10.1. Determine the specific heat per mole, h, along the co-existence curve, i.e. for a mixed system of a liquid in equilibrium with its own vapor. Treat the vapor as an ideal gas.

   Solution:

$$h = T\frac{ds}{dT}$$

$$= T\frac{\partial s}{\partial T}|_P + T\frac{\partial s}{\partial P}|_T \frac{dP}{dT}$$

$$= c_P - T\frac{\partial v}{\partial T}|_P \frac{dP}{dT} \tag{10.18}$$

where in the last line we used a Maxwell relation and c_P is the specific heat of the gas alone. Using (10.16) and $v = RT/P$ we get

$$h = c_P - \frac{q}{T} \qquad (10.19)$$

At low temperature the second term dominates and h becomes negative: extracting heat in such a way that the liquid-vapor system remains in equilibrium, the temperature rises along the co-existence curve.

Example 10.2. How does the volume of the vapor change as a function of temperature for a process proceeding along the co-existence curve?
  Solution:

$$\frac{dv}{dT} = \frac{\partial v}{\partial T}|_P + \frac{\partial v}{\partial P}|_T \frac{dP}{dT} \qquad (10.20)$$

Using again (10.16) and $v = RT/P$ we get

$$\frac{dv}{dT} = \frac{1}{P}(1 - \frac{q}{RT}) \qquad (10.21)$$

At low temperatures we have $dv/dT < 0$, so that the volume of the vapor phase decreases for increasing temperature.

10.3   Van der Waals gas

So far we have developed the thermodynamics of phase transitions in very general terms, without reference to specific models or even specific transitions, whether it is gas-solid, gas-liquid, or liquid-solid transitions in any material. In this section we want to flesh out these ideas by looking at a simple but realistic model of a fluid system. Throughout this discussion, we use the word "fluid" to encompass either a gas or a liquid.

  Recall that the ideal gas only describes a gas at low density and high temperatures, i.e. far from the transition to a liquid. An ideal gas cannot undergo a phase transition; it always remains homogeneous. In 1873, Johannes Diderik van der Waals put forward some simple ideas in his PhD thesis to Leyden University in The Netherlands that resulted in an equation of state for a "real" gas including its liquid state; he received a Nobel Prize for this work in 1910. The van der Waals equation of state reads

$$\left[P + a(\frac{N}{V})^2\right](V - Nb) = NRT \qquad (10.22)$$

  For $a = b = 0$ this obviously reduces to the ideal gas law. What are the meanings of the additional terms? The salient assumption underlying the

Table 10.1   van der Waals constants for various fluid systems. Values of a and b reported here are those that provide the best fit to experimental equation of state data.

fluid	H_2O	$Argon$	H_2	N_2	$Propane$	$Ethanol$
$a\ [MPaL^2/mol^2]$	0.5537	0.135	0.0245	0.137	0.938	0.126
$b\ [L/mol]$	0.0305	0.037	0.0305	0.0387	0.090	0.087

ideal gas law is the fact that molecules are treated as point particles. The whole volume in which the gas is confined is therefore available. However, in reality atoms and molecules have a finite, albeit very small volume, typically several $Å^3$. Recall that at room temperature and atmospheric pressure a mole of a gas ($N_A = 6.023 \times 10^{23}$ molecules) occupies about $24L$. The combined molecular volume of that number of molecules is about $10^{-4}L = 0.1cm^3$, a negligible fraction of the volume occupied by the gas. On the other hand, in the liquid state the density is much higher but still less than that of all molecules stuck together (as in a solid); the reason is that in a liquid the molecules are still free to move around. Thus van der Waals argued that the volume available for the fluid should be reduced by the volume of the constituent molecules, hence the term $-Nb$. Next, the pressure of a gas on the walls of the container is due to the change of momentum as the gas particles scatter off the wall. This is acceptable as long as the interactions between gas particles is negligible (as one assumes for an ideal gas). However, as the density increases the probability increases that any two particles are within range of their interactions. This gives rise to the overall energy density in the fluid being proportional to the square of the density. Remember, pressure can be understood as either a force per area or as energy per unit volume; thus the term $a(N/V)^2$. A list of van der Waals constants for various gases is given in the Table.

In Figure 10.7, we show a number of isotherms of a van der Waals fluid in the $P - V$ plane. For large volumes they approach the hyperbolic isotherms of the ideal gas law, $P \sim 1/V$. At low volume they are much steeper indicating a larger compressibility which is consistent with our notion that liquids are more difficult to compress than gases. Importantly for our discussion here is the fact that at low temperature ($T < T_5$) and intermediate volume the isotherms are non-monotonic, i.e. they show an intermediate maximum and minimum between which the compressibility is negative and thus thermodynamically unstable! This instability results in a first order transition, in this case condensation (upon compression).

The situation becomes more transparent when we plot isotherms of the Gibbs free energy as a function of unconstrained volume, see Figure 10.8.

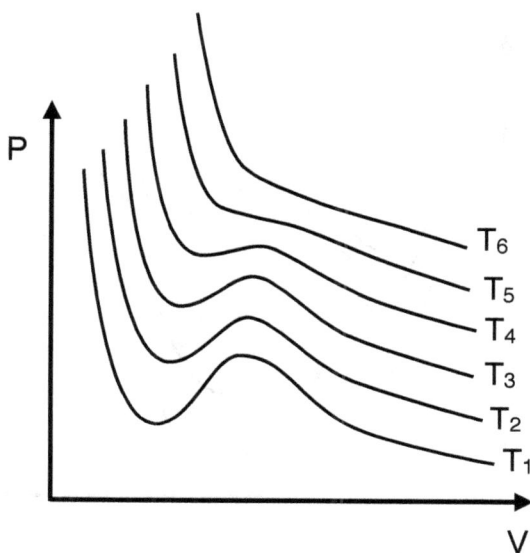

Fig. 10.7 Van der Waals isotherms. The existence of a local maxima and minima for $T_{1...4}$ implies the system is thermodynamically unstable under these conditions.

At high temperature the chemical potential is negative (ideal gas) and there is a single minimum at a molar volume corresponding to the gas phase. As temperature is lowered the Gibbs potential rises to less negative values and a second local minimum starts to develop at a much smaller molar volume corresponding to that of the liquid. As long as this local minimum is higher in energy than the gas minimum, the gas phase is stable. When the local minimum drops below that of the gas minimum the liquid state becomes the stable phase. Along the co-existence curve the two minima have the same energy indicating that both phases are stable and co-exist in equilibrium with each other; this is the curve labeled T_3.

Remark 10.3. In Figure 10.8 the volume is an unconstrained variable so that only at the absolute minimum of these curves will the system be in a stable equilibrium. Points away from the minimum can be reached if the density in a small region of the system deviates from its homogeneous average through a spontaneous fluctuation. In the gas phase this might be a region of much higher density in which a metastable liquid droplet will form. For this to happen the fluctuating region must acquire additional energy to overcome the activation barrier between the stable gaseous minimum and

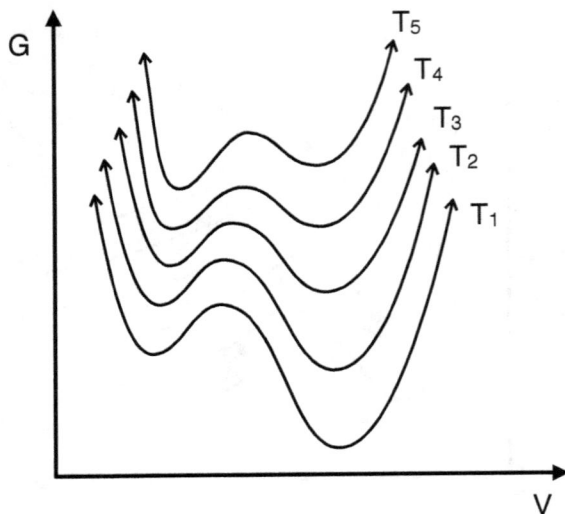

Fig. 10.8 Schematic of the Gibbs free energy as a function of the unconstrained volume for decreasing temperatures, $T_{i+1} < T_i$. The state of thermodynamic equilibrium is at the minimum of each curve. Curve labeled T_3 has two equal minima and corresponds to the co-existence region.

the metastable liquid minimum. Once in the metastable minimum, the local region must gain enough energy to overcome the barrier as seen from the local minimum. If that barrier is too high, metastability can last a long time and look like stable equilibrium. This is what happens with diamond at room temperature.

Because equality of the chemical potentials between the two phases determines the co-existence line, it is incumbent upon us to construct it from the van der Waals isotherms. To do this we start with the Gibbs-Duhem equation

$$d\mu = -sdT + vdP \qquad (10.23)$$

which we integrate at constant temperature to get

$$\mu = \int v(P)dP + f(T) \qquad (10.24)$$

which along an isotherm becomes, between two states A and B,

$$\mu_B - \mu_A = \int_A^B v(P)dP \qquad (10.25)$$

To aid the eye we re-plot an isotherm both in the $P-V$ and, equivalently, in the $V-P$ plane (see Figure 10.9).

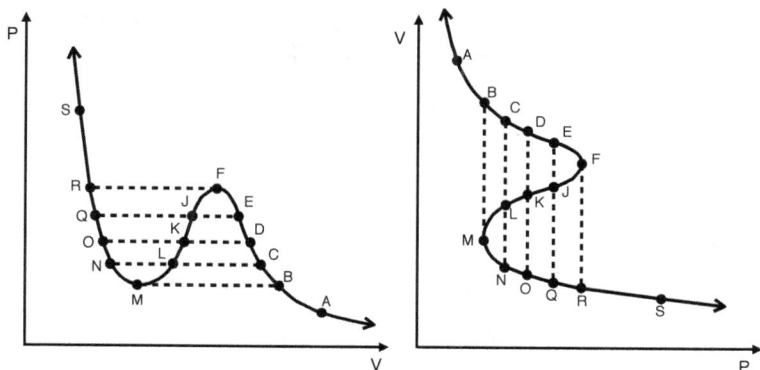

Fig. 10.9   A van der Waals isotherm in the P versus V and V versus P plane for the evaluation of the chemical potential.

Starting at low pressure and high volume in the gas phase, the chemical potential increases up to the local maximum of the van der Waals loop (point F). Because $P(V)$ subsequently decreases, so does the chemical potential up to the point M of the minimum of $P(V)$. With the latter rising now for lower volume, the contribution to the integral for the chemical potential becomes positive again, and the chemical potential increases. This produces the "butterfly" in Figure 10.10.

The thermodynamically stable portions of the butterfly are the minima, $A \rightarrow D$ for the gas phase and $O(=D) \rightarrow S$ for the liquid; the intersection of these two branches at $D = O$ signal two phase co-existence. Above the critical temperature (the upper end of the co-existence curve) the butterfly disappears.

At the point $D = O$ the chemical potentials are equal, $\mu_O = \mu_D$, or

$$0 = \mu_O - \mu_D = \int_D^O v(P)dP$$

$$= \mu_F - \mu_D + \mu_K - \mu_F + \mu_M - \mu_K + \mu_O - \mu_K \qquad (10.26)$$

or, re-arranging

$$\int_D^F v(P)dP - \int_K^F v(P)dP = \int_M^K v(P)dP - \int_M^O v(P)dP \qquad (10.27)$$

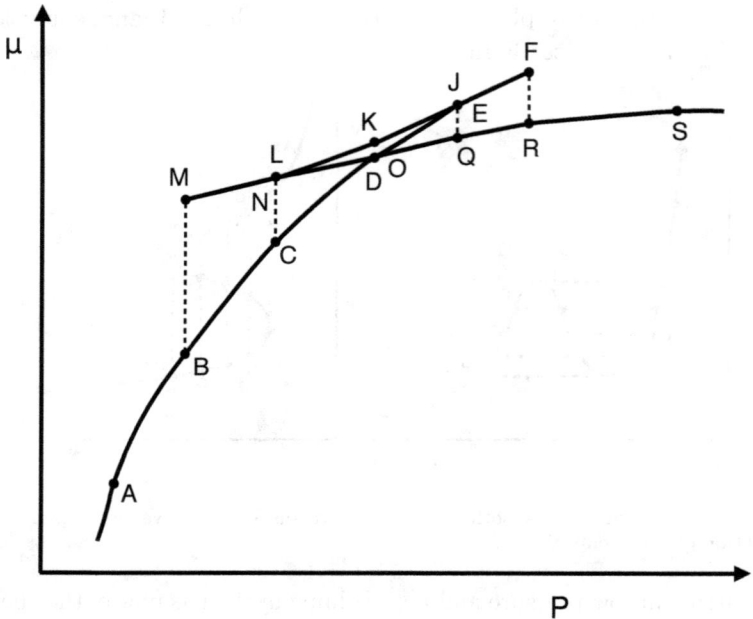

Fig. 10.10   Chemical potential versus pressure (the "butterfly") obtained by integrating a van der Waals isotherm.

A little inspection shows that the left hand side is equal to the area I of the enclosed region $DFKD$, and the right hand side is the area II of the enclosed region $KMOK$ (see Figure 10.11). Thus the area rule, or Maxwell construction,

$$\text{area } I = \text{area } II \qquad\qquad (10.28)$$

turns the "van der Waals loop" with its unstable portion FM into a thermodynamically stable equation of state, with the volume region along the straight line between D and O comprising the co-existence region at that temperature, as shown in Figure 10.11.

Repeating this construction for all van der Waals loops maps out the co-existence region as shown in Figure 10.11. Taking the slopes of the stable portions of the butterfly at point $D = O$ gives the respective entropies and their difference gives the latent heat. In addition, the lever rule (10.6) gives us the mole fractions of gas and liquid in the co-existence region.

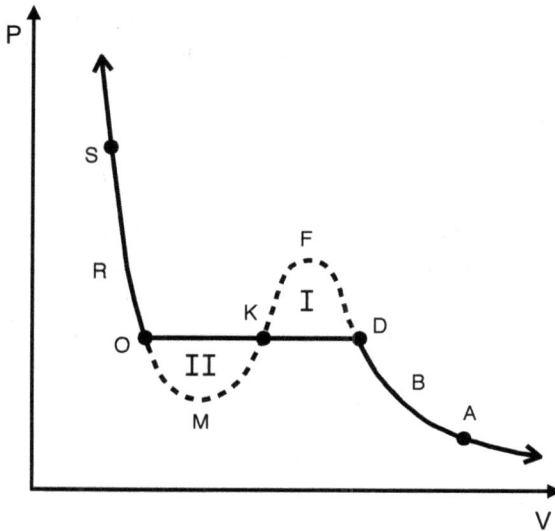

Fig. 10.11 The Maxwell construction of the stable thermodynamic isotherm. Region $D \to F$ is unstable supersaturated vapor and $M \to O$ is unstable superheated liquid.

10.4 Metastability: supersaturated water and overheated liquid

The appearance of an unstable region ($F \to M$ in Figure 10.11) of the van der Waals isotherm with a negative compressibility lead us to the Maxwell construction of a stable isotherm. In the course of doing this we also discarded the regions $D \to F$ and $M \to O$. Along the line $D \to F$ the fluid is still in the homogeneous vapor phase, but at a pressure that is higher than the vapor pressure for co-existence with the liquid at the same molar volume. This region is called supersaturated vapor or undercooled vapor because these (P,V) points actually lie on stable isotherms of a higher temperature. The undercooled vapor is thermodynamically unstable and can only be produced in very careful experiments with absolutely clean gases. Gas liquefaction requires small particles such as dust that act as nucleation seeds. This phenomenon is used in the Wilson cloud chamber in which ionized particles, produced in an elementary particle accelerator or by cosmic rays, act as nucleation seeds in supersaturated water vapor leaving behind a trail of droplets (see Figure 10.12).

What can thermodynamics say about the growth of droplets by nucleation in a supersaturated vapor? We consider a sphere of radius R that can be either filled with a gas or its corresponding liquid. The difference in Gibbs free energy for the two options is

$$\Delta G = G_l - G_g = -\frac{4\pi}{3}R^3 n_l \Delta\mu + 4\pi R^2 \sigma \tag{10.29}$$

The first term is the gain in free energy transferring $(4\pi/3)R^3 n_l$ moles from the gas phase into the volume of the droplet (assumed spherical) to a much higher liquid molar density n_l where $\Delta\mu = \mu_g - \mu_l$ is the gain in free energy per mole; $\Delta\mu > 0$ for the liquid phase to be more stable. This term favors droplets. The second term is the energy cost to establish a surface with σ, the surface tension, which is the free energy per surface area; it counteracts droplet formation. For small R, ΔG rises due to the second term. As R increases the first term dominates. The maximum in ΔG is the radius below which droplets shrink and above which they continue growing. To get this critical radius for nucleation we set the derivative of ΔG with respect to R equal to zero

$$\frac{\partial \Delta G}{\partial R} = 0 = -4\pi R^2 n_l \Delta\mu + 8\pi R\sigma \tag{10.30}$$

to get

$$R_c = \frac{2\sigma}{n_l \Delta\mu} \tag{10.31}$$

At this radius we have

$$\Delta G_c = \frac{16\pi}{3} \frac{\sigma^3}{n_l^2 (\Delta\mu)^2} \tag{10.32}$$

This amount of free energy must be provided at the nucleation centre by a thermal fluctuation, i.e. a temporary and local inhomogeneity in the vapor phase with higher free energy.

To get an estimate we treat the vapor phase as an ideal gas and get

$$\Delta\mu = RT \ln(P/P_{sat}) \tag{10.33}$$

For water we have $\sigma = 7.2 \times 10^{-2}\,\mathrm{J/m^2}$. Let us (arbitrarily) assume that the supersaturation pressure is $P = 1.1 P_{sat}$ is 10% larger that the equilibrium saturation or vapor pressure. At this pressure and at $T = 300$ K the molar density of liquid water is $n_l = 5 \times 10^4$ mol/m^3 (mass density is 10^3 kg/m^3). This gives $R_c \approx 10^{-8}$ m, i.e. a tiny droplet with about 10^5 water molecules. For this situation we have $\Delta G_c = 4 \times 10^{-18}$ J. Thus the energy in a fluctuation needed to grow the droplet beyond its critical radius (so that it can continue to grow) is about 1% of the thermal energy. This is actually a lot for a fluctuation. Consequently, the metastable supersaturated vapor is long-lived except when something colossal happens such as shaking the vessel or ionizing radiation passing through depositing much more energy than that (see Figure 10.12).

Another way in which a droplet may form is if the vapor has contaminants present, e.g. aerosols in the atmosphere. The presence of these particles provides nucleation "seeds" from which water droplets can form. There are several sources of aerosols in our atmosphere, including natural ones (e.g. sea salt), as well anthropogenic ones such as condensation trails (contrails) from jet aircraft. Higher concentrations of aerosols can result in significant increases in local cloud coverage. It is known that clouds play an important role in global temperature regulation. Whether the net effect of increased cloud coverage is a positive or negative feedback in global warming is still an open question.

Remark 10.4. Similar to the supersaturated vapor, the region $M \to O$ is also unstable and represents an overheated liquid. At those points the fluid should already exhibit co-existence. Again, a slight disturbance will cause the liquid to rapidly start boiling. This can be disastrous, e.g. when boiling water in a clean pot. To avoid overheating the water, you can put a spoon into it or place a piece of ceramic at the bottom of the pot. In cosmic ray and elementary particle physics, this phenomena is exploited in bubble chambers (a tank filled with superheated liquid hydrogen). Charged particles passing through the liquid act as nucleation seeds for bubbles, thus marking their trails.

Remark 10.5. Alloys such as bronze (copper and tin) and brass (copper and nickel) or steel also undergo first order phase transitions remarkably similar to fluids. The "gas-like" phase is complete mixing of two or more components and the "liquid-like" phase is segregation into regions of pure components. The fact that latent heat accompanies this transition can be demonstrated by a spectacular experiment. We heat a ribbon of iron with a high content of carbon, about 1% by volume, to a temperature above 1000° C (high enough so that it glows yellow) by sending an electric current through it. We then switch the current off and the the sample cools down - the glow disappears. Suddenly, as a temperature of 720° C is reached, there is a bright flash of light! At that temperature the sample undergoes a phase transition separating into pure iron and small regions of iron carbide. The sudden release of the latent heat causes the sample to temporarily heat up, which we see as a flash of light.

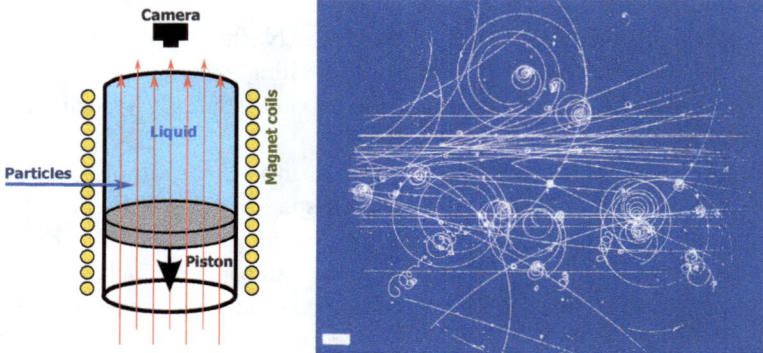

Fig. 10.12   Schematic of a bubble chamber and tracks of charged (spirals in a magnetic field) and uncharged (straight lines) elementary particles. Image credits [Wikimedia-Commons (2007c); CERN (1970)].

10.5   The critical point

As we discussed earlier, the co-existence line in the $P - T$ plane of a gas-liquid system terminates at the critical point. It is defined by a critical temperature T_{crit}, a critical pressure P_{crit}, and a critical molar volume v_{crit}. It is very important to realize that as we approach the critical point along the co-existence line, the latent heat decreases, and precisely at the critical point it vanishes. Thus at and beyond the critical point the gas-liquid transition is no longer first order.

For temperatures and pressures above their critical value there is no difference between gas and liquid and the fluid remains homogeneous. In the $T - V$ and $P - V$ planes the critical point corresponds to the maximum of the co-existence region.

This leads to a surprising phenomenon: start the system at a temperature on the right side of the co-existence region in the $P - V$ plane, see Figure 10.14. Then reduce the molar volume at constant pressure. When you hit the co-existence region the system becomes inhomogeneous with gas and liquid co-existing. As you reach the left-hand side of the co-existence region the inhomogeneity disappears and you are in the homogeneous liquid phase. Now do a second experiment: start at the same initial state but this time decrease molar volume and increase the pressure in such a way that you circumvent the critical point. When you are above but on the left of it, reduce the pressure and reduce the volume if needed until you reach the final state of the previous experiment. In this second excursion you have

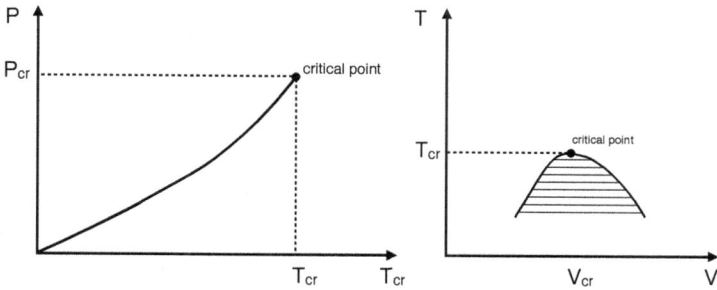

Fig. 10.13   Critical point in the $P - T$ and $T - V$ planes.

never encountered a two-phase situation because you can only identify two phases when they co-exist!

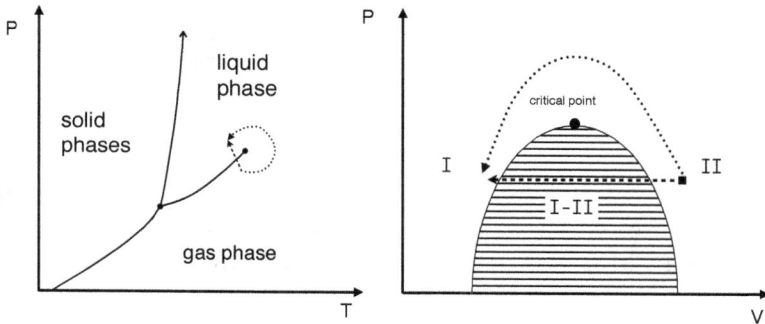

Fig. 10.14   Going through the co-existence region and around the critical point.

Having bypassed the critical point, we do another experiment to determine what happens exactly at the critical point: starting from a point on the co-existence line (in the $P-T$ plane) increase temperature and pressure so that you stay on the co-existence line. We do this with a vial of such a volume and mole number so that we actually can reach the critical point. At any stage you see an interface between the two phases. If you were boiling the liquid away, the liquid would diminish and the interface would recede until you leave the co-existence region and enter the homogeneous gas phase. Doing this along the co-existence line ends with a spectacular phenomenon called critical opalescence: as you reach the critical point the vial suddenly turns almost black. The fluid inside undergoes violent motion, starting at the interface. Eventually the interface disappears mid-air.

After the violence has ceased (you are now above the critical point), the system is completely homogeneous with no interface whatsoever.

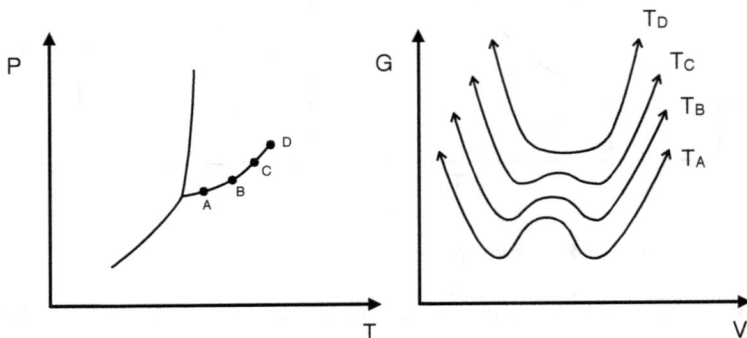

Fig. 10.15   The Gibbs free energy shows two equal minima in the co-existence region. As the critical point is reached the two minima merge into a flat-bottomed well. The fact that G is constant over a wide range of values implies that there is no thermodynamic "restoring force" to damp oscillations.

To get a qualitative idea about the new physics happening at the critical point, we follow our approach to it with a sequence of isothermal plots of the Gibbs free energy as a function of volume (see Figure 10.15). As we have seen earlier along the co-existence line, the Gibbs potential has two local minima of equal depth, one for the gas and one for the liquid, separated by an energy barrier. In both minima equilibrium is maintained because if, for example, the liquid phase were to make an excursion away from the minimum, it would be restored to it by minimizing its energy. As we approach the critical point the energy barrier between the two minima gets smaller and smaller until it completely vanishes at the critical point. The result is that now the Gibbs free energy has a flat region, where the system is stable, over a range of molar volumes (or densities). By stable we mean that all of the states have the same energy. The result is that there are no restoring forces taking the system back into a unique equilibrium state. Fluctuations are no longer damped but become macroscopic. In a fluid, light scattering becomes so large that the system looks black! The further result of this is that thermodynamic extensive variables undergo fluctuations that are of the order of the quantities themselves and are no longer small negligible corrections. New physics and mathematics are needed: the theory of critical phenomena. This is one area of physics where discontinuous functions and singularities appear.

10.6 The law of corresponding states and universality

Our brief discussion of critical phenomena suggest that an analytic model such as van der Waals' can be at most qualitatively correct, and this is indeed the case. Still, one can gain some insights. We therefore proceed to calculate the critical parameters T_{crit}, V_{crit}, and P_{crit}, as van der Waals would predict them.

At the critical temperature, isotherms have an inflection point for which both the first and second derivatives vanish. For the van der Waals model this happens when

$$\frac{\partial P}{\partial V}|_T = -\frac{NRT}{(v - Nb)^2} + \frac{2N^2a}{V^3} = 0$$

$$\frac{\partial^2 P}{\partial V^2}|_T = \frac{2NRT}{(v - Nb)^3} - \frac{2N^2a}{V^4} = 0 \qquad (10.34)$$

which, together with the van der Waals equation allows us to obtain the critical values

$$T_{crit} = \frac{8}{27}\frac{a}{b}$$

$$V_{crit} = 3Nb$$

$$P_{crit} = \frac{1}{27}\frac{a}{b^2} \qquad (10.35)$$

in terms of the van der Waals constants. As the critical values are of course available from experiment, one may want to turn the argument around by rescaling temperature, pressure, and density

$$\overline{T} = T/T_{crit}$$

$$\overline{P} = P/P_{crit}$$

$$\overline{V} = V/V_{crit} \qquad (10.36)$$

resulting in the van der Waals equation in dimensionless form

$$(\overline{P} + \frac{3}{\overline{V}^2})(3\overline{V} - 1) = 8\overline{T} \qquad (10.37)$$

valid for all systems for which the van der Waals model is qualitatively acceptable. The states of two different fluid systems for which the values of the reduced variables are the same are called corresponding states. The $P-V$ isotherms for corresponding states are the same, as is the co-existence

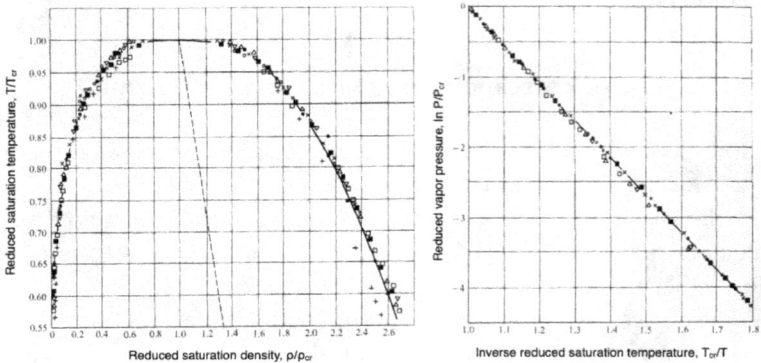

Fig. 10.16   Vapor pressure curves in reduced variables $T' = T/T_{\text{crit}}$, $P' = P/P_{\text{crit}}$, $v' = v/v_{\text{crit}}$. Different symbols are for different gases: Ne, Ar, Xe, N_2, O_2, CO, CH_4.

region. In addition, the following quantities are the same: (1) the reduced vapor pressure, (2) the reduced molar volume of the vapor, (3) the reduced molar volume of the liquid in equilibrium with its vapor.

For simple systems the law of corresponding states is satisfied amazingly well, despite the fact that it has been derived from a theory that is only qualitatively correct at best. In Figure 10.16, we demonstrate this point for a number of gases.

In this chapter we have mostly talked about the first order transition in the fluid system. The derivation of the latent heat from equality of chemical potentials of co-existing phases, and the derivation of the Clausius-Clapeyron equation are very general. These ideas apply to any first order transition no matter what the system is. To illustrate this point, in the table below we have listed a number of systems exhibiting first order phase transitions. Because these transitions show a universal behavior, the theory should be formulated in terms of a variable that in different systems has a different physical realization. L.D. Landau introduced such a variable in 1937 and called it the order parameter. Landau's theory of phase transitions is based on a phenomenological Gibbs free energy in which the order parameter is the crucial intensive variable in addition to T and P.

System	discontinuity	order parameter
liquid - vapor	molar volume	$v_{vapor} - v_{liquid}$
liquid - solid	molar volume	$v_{liquid} - v_{solid}$
solid - solid	symmetry	$v_{solid(1)} - v_{solid(2)}$
binary alloys	de-mixing	$v_{mixed} - v_{de-mixed}$
partially miscible fluids	de-mixing	$v_{mixed} - v_{de-mixed}$
ferromagnet	magnetization	spontaneous magnetization
ferroelectric	electric dipole	dipole moment
superconductivity	electric conductivity	energy gap
superfluidity	entropy	macroscopic wave function

10.7 Landau theory of phase transitions

In this section we give a brief account of Landau's theory of phase transitions. We recall that the gas-liquid transition is accompanied by latent heat and that within the co-existence region the Gibbs free energy has two minima as a function of volume (see Figure 10.15). One of these minima corresponds to the molar volume of the vapor phase, v_{vapor}, while the other is the much smaller molar volume of the liquid, v_{liquid}.

As one travels along the co-existence line, the difference in the two volumes (which Landau identified as the order parameter in the gas-liquid transition), must decrease and eventually vanishes at the critical point (as does the latent heat). As a result, the phase transition at the critical point becomes second order. To describe the behavior of the Gibbs free energy along the co-existence line close to the critical point (where the order parameter is small), Landau postulated that as a function of the order parameter the Gibbs free energy can be expanded as

$$G = G_0 + G_2(T)\psi^2 + G_4(T)\psi^4 + ... \qquad (10.38)$$

Remember that along the co-existence line the pressure is a function of temperature $P = P(T)$. To ensure a single minimum above T_{crit} we need $G_4 > 0$ and $G_2(T > T_{\text{crit}} > 0)$. To get two minima below T_{crit}, we need $G_2(T < T_{\text{crit}}) < 0$. We therefore use an expansion

$$G_2(T) = g_2(T - T_{\text{crit}}) + ... \qquad (10.39)$$

This is the case for the van der Waals gas, see Problem 10.2. At the critical point $\psi = 0$ and

$$\frac{\partial G}{\partial \psi} = 2g_2(T - T_{\text{crit}}) + 4G_4\psi^3 + ... = 0 \qquad (10.40)$$

Below T_{crit} there are two solutions

$$\psi = \pm \left[2\frac{g_2}{G_4}(T_{\text{crit}} - T)\right]^{\beta} \qquad (10.41)$$

with $\beta = 1/2$. Again, this is the case for the van der Waals gas for which

$$v_g - v_l = D(T_{\text{crit}} - T)^{1/2} \qquad (10.42)$$

It is left to Problem 10.3 to determine the coefficient D. Experiments on a range of gases, on alloys and on magnetic systems show that $\beta \approx 0.35$ instead, signaling that an analytic theory as postulated by Landau and resulting from thermodynamic arguments is close but not correct! Continuing with the van der Waals gas (rather than the general Landau theory) one can show that

$$P - P_{\text{crit}} = -A(T - T_{\text{crit}})(v - v_{\text{crit}}) - \frac{1}{3}B(v - v_{\text{crit}})^3 + \dots . \qquad (10.43)$$

Experiments on gases show that at $T = T_{\text{crit}}$

$$|P - P_{\text{crit}}| \sim |v - v_{\text{crit}}|^{\delta} \qquad (10.44)$$

with $\delta \approx 4.4$, instead of the Landau/van der Waals prediction $\delta = 3$. From (10.43) we also get the compressibility

$$\kappa_T(v = v_{\text{crit}}) \sim |T - T_{\text{crit}}|^{-\gamma} \qquad (10.45)$$

with the Landau/van der Waals prediction $\gamma = 1$ whereas experiments on a range of gases show $\gamma \sim 1.26$.

Lastly, we mention that the specific heat of a van der Waals gas along an isochore in the two-phase region rises linearly to a value $c_V = 6R$ at T_{crit}, and has a value $3/2R$ above T_{crit}. This is in complete disagreement with experimental data which show that

$$c_V \sim |T - T_{\text{crit}}|^{-\alpha} \qquad (10.46)$$

is symmetrical around T_{crit} with $\alpha \approx 0.04$.

Thermodynamic stability arguments, similar to the ones used in Chapter 9, imply the following relations for the critical exponents

$$\beta(\delta + 1) \geq 2 - \alpha \qquad (10.47)$$

$$\gamma + 2\beta \geq 2 - \alpha \qquad (10.48)$$

Experimental data confirm that these two relations are almost equalities.

Remark 10.6. The discrepancies in the critical exponents between experimental results and values obtained from an analytical form of the Gibbs free energy and thus of the equations of state are conclusive arguments that physics at the critical point is non-analytical: we have reached the limitation of thermodynamics! Why? And what is the physical origin of this discrepancy?

Thermodynamics is based on the premise that macroscopic objects can be characterized by a small number of macroscopic variables. This is definitely possible if the fluctuations around these averaged values of volume, mole number, energy, etc. are negligible. However, experiments on critical opalescence show that light scattering becomes enormous as the critical point is approached, and density fluctuations reach macroscopic proportions because the Gibbs free energy attains a flat bottom without any restoring forces. Thus thermodynamics has reached its limit of applicability. What is needed is a theory that incorporates fluctuations: this is done in statistical mechanics. To understand how this is done, we rephrase the arguments by stating that fluctuations are the manifestation of correlated motion. If such correlations are short-ranged, fluctuations are microscopically small and short-lived. On the other hand, fluctuations over macroscopic distance, i.e. over the whole volume of the system, are the result of macroscopically large correlations. It took almost fifty years after Landau's analytical theory of critical phenomena and one hundred years after van der Waals before the modern theory of critical phenomena was developed. This culminated in the Renormalization Theory of Kenneth G. Wilson in 1971 (Nobel prize 1982) that accounts rigorously and effectively for correlations.

10.8 No phase transitions in one-dimensional systems (almost)

Before we start this discussion we should be specific of what we mean by one-dimensional systems. Recalling Section 5.5, a polymer molecule is a linear assembly of monomers that can be curled up in three-dimensional space, but its internal dynamics are restricted to one dimension. Similarly, consider a crystal surface with deep grooves running in straight lines; the (110) surface of a bcc crystal is a good example. When this surface is exposed to a gas, molecules will preferentially adsorb either in the grooves or along the ridges between grooves. Adsorbed molecules have limited mobility perpendicular to the grooves or ridges. Such an adsorbate is obviously one-dimensional.

In some polymers, the monomers can be in one of two states which differ in length. One example are polysaccharides (sugars) where the two states are called boat- and chair-like. Another example is DNA, in which the subunits can be double-stranded (the two strands are curled around each other forming the famous double helix), or they can be uncurled with the two strands parallel to each other but still bound together. In the latter configuration the subunit is much longer. As the DNA is stretched, for instance by shear forces in blood flow, more and more subunits are converted into the longer state. We can therefore view the DNA molecule as a one-dimensional system in which short and long subunits are somehow distributed along the chain. Likewise, at finite coverage an adsorption site along the groove is either occupied or unoccupied, i.e. we again have a distribution of two states.

In general, in a one-dimensional system there will be stretches or domains of A-states (short monomers, unoccupied states in adsorption) and domains of B-states (long monomers, occupied states). In one-dimensional adsorption such domains of occupied states will be the result of attractive interactions between neighboring adsorbed molecules. In DNA domains of double-stranded DNA are due to the fact that they form the energetically most favorable configuration (as long as no forces are applied).

Continuing with the general picture we assume that there are n junction points between domains of A-states and B-states, called domain walls. Each such domain wall is associated with an energy loss ε (it is a loss because one of the two states is energetically favorable, implying that introducing a B-state costs energy). Ignoring all other interactions, the internal energy is thus $U = n\varepsilon$. Next, the distribution of n domain walls on N sites generates entropy (5.76)

$$S/k_b = N[n/N \ln(n/N) - (1 - n/N)\ln(1 - n/N)] \qquad (10.49)$$

which for $n/N \ll 1$, i.e. large domains or few domain walls, becomes

$$S/k_b \simeq N[n/N \ln(n/N) - 1] \qquad (10.50)$$

so that the Gibbs free energy becomes

$$G = G_0 + U - TS$$
$$= G_0 + n\varepsilon + Nk_BT[n/N \ln(n/N) - 1] \qquad (10.51)$$

and its derivative is

$$\frac{\partial G}{\partial n} = \varepsilon + k_BT \ln[n/N] \qquad (10.52)$$

For $n/N \ll 1$ we have $\partial G / \partial n < 0$ so that G decreases with increasing n. On the other hand, G will assume a minimum in equilibrium, i.e. n will increase to the point where it is of order N as long as the the derivative remains negative. Thus the domains will decrease in size and phase separation is not possible!

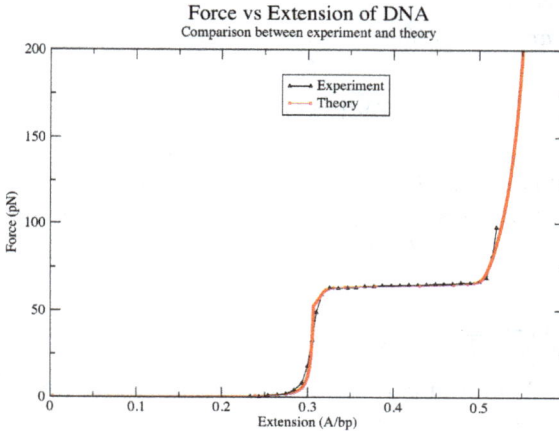

Fig. 10.17   Stretching a single DNA molecule. In the "plateau" region the double helix is unwound. This is NOT a phase transition.

As an example, we show the mechanical equation of state of DNA (its force-extension curve; pressure in the one-dimensional system is the negative force needed to stretch the molecule). Although the appearance of this curve is very much like the physical van der Waals isotherm, there is a big difference, namely the "plateau" in the force extension curve has a finite albeit very small slope, thus it is not a signature for phase co-existence.

Remark 10.7. It must be stressed that phase transitions are possible in one-dimensional systems if long-ranged interactions are present that would, as an example, act between two distant domain walls. Another possibility would be if some interaction energy were infinite, for instance the energy ε, because then the derivative of the Gibbs free energy would then always be positive. However, these are largely artificial situations.

10.9 Problems

Problem 10.1. *Derive the critical values T_{crit}, P_{crit}, and V_{crit} for the van der Waals gas and write the latter in dimensionless form.*

Problem 10.2. *Derive equations for the boundary of the two-phase region for a van der Waals gas. Solve these equations approximately in the vicinity of the critical point.*

Hint: The area rule (10.28) can be written as

$$\int_{V_l}^{V_g} P(V)dV = P(V_l)\,(V_g - V_l)$$

where the right hand side is the integral over the van der Waals loop and the left hand side is the same along the physical isotherm. Write this equation in dimensionless form and get

$$\ln(3\overline{V_g} - 1) + \frac{9}{4\overline{T}}\frac{1}{\overline{V_g}} - \frac{1}{3\overline{V_g} - 1} = \ln(3\overline{V_l} - 1) + \frac{9}{4\overline{T}}\frac{1}{\overline{V_l}} - \frac{1}{3\overline{V_l} - 1}$$

Together with the van der Waals equation itself this gives $\overline{V_g}$, $\overline{V_l}$, and $\overline{P}$ as a function of $\overline{T}$.

Problem 10.3.

(a) Derive the expansion

$$P - P_{\text{crit}} = -A(T - T_{\text{crit}})(v - v_{\text{crit}}) - \frac{1}{3}B(v - v_{\text{crit}})^3 + \ldots.$$

for the van der Waals gas expressing the coefficients A and B in terms of the van der Waals constants.

(b) Calculate the isothermal compressibility

$$\kappa_T(v = v_{crit}) \sim |T - T_{crit}|^{-\gamma}$$

determining the coefficient of proportionality.

(c) Express your results in terms of dimensionless variables.

Problem 10.4. *In an experiment, table salt is added to a beaker of water (at room temperature) until the liquid becomes saturated and any additional salt settles at the bottom of the jar. The liquid is then heated, increasing the solubility of the water thereby allowing more salt to dissolve. In the next step, the beaker is allowed to cool back to room temperature. Now there are no salt crystals at the bottom of the beaker, and the liquid is said to be super-saturated.*

(a) *How do you know that this final state is not the equilibrium one?*

(b) *If a small amount of additional salt is added to the super-saturated liquid (or it is disturbed in some other way), the system will undergo a rapid transition. A crystal will begin to grow out of the liquid. As it does this, the system will release heat. What is the origin of this heat?*

(c) *If during the cooling phase the beaker was cooled to a temperature lower than room temperature (say 10° C) will there be more or less heat released than in b). Why?*

Problem 10.5. *Based on the figure below, design an experiment to determine the latent heat of a substance. State explicitly, and justify, any assumptions your method is based upon.*

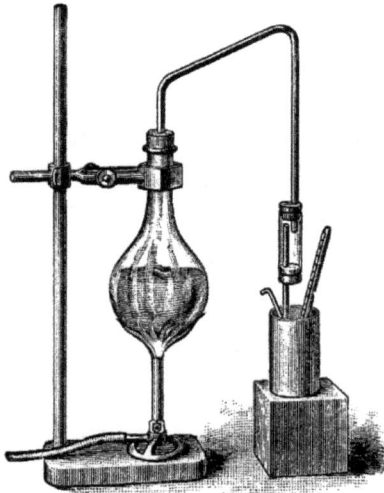

Fig. 10.18   Measuring the latent heat of vaporization. Image credit [Carhart and Chute (1912)].

Chapter Summary

(1) Systems in equilibrium are inhomogeneous when two or more phases of different densities co-exist.

(2) The co-existence of two phases is possible along the co-existence line $P = P(T)$. Its terminal points are the critical point and the triple point.

(3) In a first order phase transition, first derivatives of the Gibbs free energy such as entropy and molar volume change discontinuously. They are accompanied by a latent heat.

(4) The Clausius-Clapeyron equation controls $P(T)$ along the co-existence line.

(5) The van der Waals equation is a qualitative description of the liquid-vapor transition.

(6) Scaling temperature, pressure, and molar volume with their critical values for a given system produces the Law of Corresponding States.

(7) Thermodynamics cannot explain critical exponents.

Chapter 11

Summary of Useful Results and Final Remarks

11.1 Thermodynamic potentials

11.1.1 *Entropy $S(U, V, N)$*

Entropy is a monotonic function of the internal energy U for fixed V and N. It must be differentiable and extensive.

$$dS = \frac{1}{T}dU + \frac{P}{T}dV - \frac{\mu}{T}dN$$

11.1.2 *Internal energy and others*

Euler relation $U = TS - PV + \mu N$

$$dU = TdS - PdV + \mu dN$$
$$F(T, V, N) = U - TS$$
$$dF = -SdT - PdV + \mu dN$$
$$H(S, P, N) = U + PV$$
$$dH = TdS + VdP + \mu dN$$
$$G(T, P, N) = U - TS + PV = N\mu$$
$$dG = -SdT + VdP + \mu dN$$

Useful formulae

$$U = -T^2 \frac{\partial(F/T)}{\partial T}|V$$
$$H = -T^2 \frac{\partial(G/T)}{\partial T}|P$$

Gibbs-Duhem equation for one-component system

$$d\mu = -sdT + vdP$$

Some "greek":

- **Isobaric** means a process at constant pressure.
- **Isochoric, or isometric/isovolumetric**, means a process at constant volume.
- **Isothermal** means a constant temperature process.
- **Adiabatic** means a process that occurs without the gain or loss of energy by heat.
- **Isentropic** means a **reversible** adiabatic process, i.e. one that occurs at a constant entropy.
- **Isenthalpic** means a process which occurs at a constant enthalpy.

11.2   Ideal gas

$$PV = NRT$$
$$U = cNRT$$
$$c = c_V/R$$
$$monatomic\ gas\ c = 3/2$$
$$air\ c = 2.44$$
$$c_P = c_V + R$$
$$adiabat\ PV^\gamma = const$$
$$\gamma = c_P/c_V$$

$$S(U,V,N) = Ns_0 + NR\ln\left[\left(\frac{U}{U_0}\right)^c \frac{V}{V_0}\left(\frac{N}{N_0}\right)^{-(c+1)}\right]$$

$$U(S,V,N) = U_0\left(\frac{V_0}{V}\right)^{1/c}\left(\frac{N}{N_0}\right)^{(c+1)/c}\exp\left[S/(cNR)\right]$$

$$F(T,V,N) = NRT\left\{\frac{F_0}{N_0RT_0} - \ln\left[\left(\frac{T}{T_0}\right)^c \frac{V}{V_0}\frac{N_0}{N}\right]\right\}$$

$$H(S,P,N) = (c+1)PV = \frac{c+1}{c}H_0 P^{1/(c+1)}\frac{N}{N_0}\exp\left[(1+1/c)S/(NR)\right]$$

$$G(T,P,N) = N\mu(T,P) = N\mu_0(T_0,P_0) - (c+1)NRT\ln(T/T_0) + NRT\ln(P$$

isobaric P=const.	isochoric V=const.	isothermal T=const.	adiabatic S=const.
$cP(V_f - V_i)$	$Nc_V(T_f - T_i)$	0	$[P_f V_f - P_i V_i]/(\gamma - 1)$
$P(V_i - V_f)$	0	$Nc_V \ln(V_i/V_f)$	ΔU
$(c+1)P(V_f - V_i)$	ΔU	$-W_{i \to f}$	0

Mixture of ideal gases

$$S = \sum_j N_j s_{j0} + \sum_j N_j c_j R \ln(T/T_0) + NR \ln(V/Nv_0) + \Delta S_{mixing}$$

$$entropy \ of \ mixing \ \Delta S_{mixing} = -R \sum_j N_j \ln(N_j/N)$$

11.3 Van der Waals gas

$$\left[P + a(\frac{N}{V})^2\right](V - Nb) = NRT$$

Critical values

$$T_{crit} = \frac{8}{27}\frac{a}{b}$$
$$V_{crit} = 3Nb$$
$$P_{crit} = \frac{1}{27}\frac{a}{b^2}$$

rescaled temperature, pressure and density

$$\overline{T} = T/T_{crit}$$
$$\overline{P} = P/P_{crit}$$
$$\overline{V} = V/V_{crit}$$

Van der Waals equation in dimensionless form: Corresponding states

$$(\overline{P} + \frac{3}{\overline{V}'^2})(3\overline{V} - 1) = 8\overline{T}$$

11.4   Polymers

Wormlike Chain Model:

$$f = \frac{k_B T}{L_p(T)} \left[\frac{L_c}{4(1 - L/L_c)^2} - \frac{1}{4} + \frac{L}{L_c} \right]$$

L_c contour length of the polymer; L_p its persistence length.

11.5   Joule-Thomson throttling

$$dT = \frac{v}{c_P}(T\alpha - 1)dP$$

11.6   Engines

Carnot efficiency: dW_{RWS} work delivered to the work source and heat $(-dQ_h)$ extracted from the hot reservoir

$$\varepsilon_{Carnot} = \frac{dW_{RWS}}{(-dQ_h)} = 1 - \frac{T_c}{T_h}$$

Otto cycle efficiency: $\gamma = c_P/c_V$ ratio of specific heats; $r = V_A/V_B$ compression ratio

$$\varepsilon_{Otto} = 1 - \left(\frac{V_B}{V_A}\right)^{\gamma-1}$$

$$= 1 - \frac{1}{r^{\gamma-1}}$$

Diesel cycle efficiency: $r_c = V_A/V_C$ cut-off ratio

$$\varepsilon_{Diesel} = 1 - \frac{1}{r^{\gamma-1}} \frac{r_c^{\gamma} - 1}{\gamma(r_c - 1)}$$

Refrigerator efficiency: $(-dQ_c)$ heat extracted from fridge; $(-dW_{RWS})$ work used to run the fridge

$$\varepsilon_r = \frac{(-dQ_c)}{(-dW_{RWS})} = \frac{T_c}{T_h - T_c}$$

Heat pump: dQ_h heat delivered and $(-dW_{RWS})$ work bought to run the heat pump

$$\varepsilon_p = \frac{dQ_h}{(-dW_{RWS})} = \frac{T_h}{T_h - T_c}$$

11.7 Phase transitions

Latent heat

$$q = T(s_2 - s_1)$$

Clausius-Clapeyron

$$\frac{dT}{dP} = \frac{T(v_2 - v_1)}{q}$$

Clapeyron

$$\frac{dP}{dT} = \frac{qP}{RT^2}$$

$$\frac{d \ln P}{dT} = \frac{q}{RT^2}$$

11.8 Final remarks: beyond equilibrium thermodynamics

Thermodynamics deals with macroscopic systems in and close to equilibrium. A small number of macroscopic and extensive variables are needed. The entropy accounts for the internal degrees of freedom of a system and is a measure for its disorder. It is an analytic function of the extensive variables and as such it contains all the thermodynamic information about a given system. Intensive variables such as temperature, pressure, and chemical potential are obtained as derivatives of the fundamental relation.

Macroscopic implies that fluctuations are ignored. This is justified except near critical points where macroscopic fluctuations dominate the physics and non-analytic behavior in the form of singularities appears, necessitating a **statistical theory of critical phenomena.**

Thermodynamics is a phenomenological theory that needs input from experiment in the form of thermal and mechanical equations of state, or the pressure and temperature dependence of expansion coefficient, compressibility and specific heat, etc. Thermodynamics provides a multitude of relations with which the consistency of measurements can be checked and other properties can be calculated.

To make the connection with the underlying microscopic physics – classical and quantum mechanics – **statistical mechanics** has been developed that allows us to calculate fundamental relations such as the internal energy and free energies from which equations of state and other properties can be obtained in simple models. Thus an understanding of particular properties such as specific heat or compressibility can be obtained and checked against experimental data. Statistical Mechanics must comply with all the laws and rules of Thermodynamics.

Thermodynamics can deal with processes as long as they are approximately quasi-static, i.e. slow on the time scale of internal relaxation. Thermodynamics sets an upper limit on the efficiency of any cyclical process in terms of the temperatures of the two heat reservoirs to which the device is coupled.

To deal with systems out of equilibrium, thermodynamics has been generalized to **non-equilibrium thermodynamics**: one introduces local time dependent densities such as the mass density $\rho(\mathbf{r}, t)$, internal energy density $u(\mathbf{r}, t)$ and a velocity field $\mathbf{v}(\mathbf{r}, t)$. Integrating these quantities over the system results in the macroscopic variables. These quantities must satisfy balance equations such as the continuity equation for the mass density

$$\frac{\partial \rho}{\partial t} + div(\rho \mathbf{v}) = 0$$

To introduce the entropy, one must assume that locally the system is maintained in equilibrium over a volume small compared to the system, but large enough to apply thermodynamics. Thus a local entropy density can be introduced for which a balance equation can be derived. In this balance, entropy production (per unit time) plays the dominant role, as important as the second law in equilibrium thermodynamics. The set of balance equations for density, energy, momentum, and entropy controls the dynamics of processes such as diffusion, energy transport, heat conduction, friction, and many more. If such processes happen while local equilibrium is maintained, one can approximate the transport currents by linear laws, for instance according to Fourier's law the heat current is linear in its driving "force" the temperature gradient, $\mathbf{j}_q = -\lambda \nabla T$ and according to Fick's law diffusion is linear in the density gradient, $\mathbf{j}_m = -D \nabla \rho$. In this linear regime it can be shown that the entropy production is minimal for stationary (or time independent) processes. Thus running a factory such as an oil refinery in a steady state not only makes it safe (fluctuations are small) but also produces the least amount of entropy for a given throughput, i.e. pollution is minimal! The skill of a good process engineer is to maximize the output but still minimizing the entropy production. If the process is pushed too far, large fluctuations set in that lead to instabilities and possible catastrophic failure. How these ideas have implications in financial markets as well is all too well known: in times of rapid growth in the stock market, daily variations in stock prices are huge, inevitably leading to a crash sooner or later. On the other hand, in times of slow growth daily stock fluctuations are small and investment risks are low.

Returning to the physical sciences, if local equilibrium is not guaranteed one must resort to **non-equilibrium statistical mechanics** to derive kinetic equations, such as the Boltzmann equation for dilute gases, Markovian master equations, Langevin equations, and the like. They will, in particular, serve to calculate transport coefficients such as heat and electrical conductivity, friction coefficient, and diffusion coefficient to name but a few.

Appendix A

Partial Derivatives and Differential Forms

In this appendix we have collected the basic definitions and properties of partial derivatives and differential forms to the extent they are relevant in thermodynamics.

Given a function $f(x, y, z)$ of three or more variables that is piecewise continuous and differentiable (with respect to all variables), the Taylor expansion for an infinitesimal displacement is

$$f(x + dx, y + dy, z + dz) = f(x, y, z) + df + \frac{1}{2!}d^2 f + \dots \qquad (A.1)$$

where

$$df = \frac{\partial f}{\partial x}|_{y,z}dx + \frac{\partial f}{\partial y}|_{x,z}dy + \frac{\partial f}{\partial z}|_{x,y}dz \qquad (A.2)$$

is the first order differential or differential form. The partial derivatives are equivalent to ordinary derivatives because the other arguments of f are held fixed. Likewise, $d^2 f$ is a second order differential and is given in terms of second partial derivatives

$$d^2 f = \frac{\partial^2 f}{\partial x^2}|_{y,z}(dx)^2 + \frac{\partial^2 f}{\partial y^2}|_{x,z}(dy)^2 + \frac{\partial^2 f}{\partial z^2}|_{x,y}(dz)^2$$

$$+2\frac{\partial^2 f}{\partial x \partial y}dxdy + 2\frac{\partial^2 f}{\partial x \partial z}dxdz + 2\frac{\partial^2 f}{\partial y \partial z}dydz \qquad (A.3)$$

Here we have used the fact that the mixed derivatives are symmetric

$$\frac{\partial}{\partial x}\left[\frac{\partial f}{\partial y}|_{x,z}\right]_{y,z} = \frac{\partial}{\partial y}\left[\frac{\partial f}{\partial x}|_{y,z}\right]_{x,z} \qquad (A.4)$$

It is absolutely crucial to keep track of which variables are held constant in a partial derivative! Most mistakes in thermodynamics arise from ignoring this rule.

It often occurs that the variables of a thermodynamic function them-selves depend on a parameter. Take, as an example, the entropy function: its natural variables are internal energy U, volume V, and mole number N. For the ideal gas we know $U = 3/2RT$ and $V = NRT/P$ as functions of temperature. So we cannot vary these variables independently but must relate their variation to changes in T. In a more general framework let us assume that $x = x(t)$, $y = y(t)$, and $z = z(t)$. We then have

$$dx = \frac{dx}{dt}dt$$

$$dy = \frac{dy}{dt}dt \qquad \text{(A.5)}$$

$$dz = \frac{dz}{dt}dt$$

and the differential of f becomes

$$df = \left[\frac{\partial f}{\partial x}|_{y,z}\frac{dx}{dt} + \frac{\partial f}{\partial y}|_{x,z}\frac{dy}{dt} + \frac{\partial f}{\partial z}|_{x,y}\frac{dz}{dt}\right]dt \qquad \text{(A.6)}$$

Writing this as

$$\frac{df}{dt} = \frac{\partial f}{\partial x}|_{y,z}\frac{dx}{dt} + \frac{\partial f}{\partial y}|_{x,z}\frac{dy}{dt} + \frac{\partial f}{\partial z}|_{x,y}\frac{dz}{dt} \qquad \text{(A.7)}$$

we recognize that this is nothing but the chain rule.

It often happens in thermodynamics (in physics generally as well as geometry) that a constraint must be imposed on a function; this results in an implicit function

$$f(x, y, z) = constant \qquad \text{(A.8)}$$

This constraint turns one of the variables into a function of the other two, e.g. $z = z(x, y)$. The differential is identically zero

$$df = \frac{\partial f}{\partial x}|_{y,z}dx + \frac{\partial f}{\partial y}|_{x,z}dy + \frac{\partial f}{\partial z}|_{x,y}dz = 0 \qquad \text{(A.9)}$$

Because the infinitesimal changes are independent of each other we can look at the special case that $dz = 0$ and get after dividing by dx

$$0 = \frac{\partial f}{\partial x}|_{y,z} + \frac{\partial f}{\partial y}|_{x,z}\frac{\partial y}{\partial x}|_{f,z} \qquad \text{(A.10)}$$

or

$$\frac{\partial y}{\partial x}|_{f,z} = -\frac{(\partial f/\partial x)_{y,z}}{(\partial f/\partial y)_{x,z}} \qquad \text{(A.11)}$$

and similarly setting $dy = 0$ or $dz = 0$ we get, respectively

$$\frac{\partial z}{\partial x}\Big|_{f,y} = -\frac{(\partial f/\partial x)_{y,z}}{(\partial f/\partial z)_{x,y}} \tag{A.12}$$

$$\frac{\partial z}{\partial y}\Big|_{f,x} = -\frac{(\partial f/\partial y)_{x,z}}{(\partial f/\partial z)_{x,y}} \tag{A.13}$$

Looking again at the case where $dz = 0$ we divide this time by dy instead of dx and get

$$\frac{dx}{dy}\Big|_{f,z} = -\frac{(\partial f/\partial y)_{x,z}}{(\partial f/\partial x)_{y,z}} \tag{A.14}$$

Comparing this with (A.11) we get another useful result

$$\frac{\partial x}{\partial y}\Big|_{f,z} = \frac{1}{\frac{\partial y}{\partial x}\Big|_{f,z}} \tag{A.15}$$

and combining all three

$$\frac{\partial x}{\partial y}\Big|_{f,z}\frac{\partial y}{\partial z}\Big|_{f,x}\frac{\partial z}{\partial x}\Big|_{f,y} = -1 \tag{A.16}$$

Returning to (A.7) but for the case that f is constant and z is not a function of the parameter t implying that x and y are functions of t and f

$$0 = \frac{\partial f}{\partial x}\Big|_{y,z}\frac{\partial x}{\partial t}\Big|_{f} + \frac{\partial f}{\partial y}\Big|_{x,z}\frac{\partial y}{\partial t}\Big|_{f} \tag{A.17}$$

which we can write as

$$\frac{(\partial y/\partial t)_{f,z}}{(\partial x/\partial t)_{f,z}} = -\frac{(\partial f/\partial x)_{y,z}}{(\partial f/\partial y)_{x,z}} \tag{A.18}$$

Comparison with (A.11) shows that

$$\frac{\partial y}{\partial x}\Big|_{f,z} = \frac{(\partial y/\partial t)_{f,z}}{(\partial x/\partial t)_{f,z}} \tag{A.19}$$

All of the formulae collected here are useful in thermodynamics.

Bibliography

Callen, H. (1985). *Thermodynamics and an Introduction to Thermostatistics, 2nd Edition* (John Wiley and Sons, Inc., United States).

Carhart, H. and Chute, H. (1912). *First Principles of Physics* (Allyn and Bacon, United States).

CERN (1970). http://cdsweb.cern.ch/record/39474/files/11465.jpeg.

Cordes, W. (2009). http://commons.wikimedia.org/wiki/File: Vapor_Pressure_of_Water.png.

Emden, R. (1938). Why do we have winter heating? *Nature* **141**, pp. 908–909.

WikimediaCommons (2005). http://commons.wikimedia.org/wiki/File: Carbon_basic_phase_diagram.png.

WikimediaCommons (2006). http://commons.wikimedia.org/wiki/File: BlackbodySpectrum_loglog_150dpi_en.png.

WikimediaCommons (2007a). http://commons.wikimedia.org/wiki/File:Solar_Spectrum.png.

WikimediaCommons (2007b). http://commons.wikimedia.org/wiki/File:Feynman%27s_ratchet.jpg.

WikimediaCommons (2007c). http://commons.wikimedia.org/wiki/File:Bubble-chamber.svg.

WikimediaCommons (2008). http://commons.wikimedia.org/wiki/File:Feynman_ratchet.jpg.

Zureks (2006). http://commons.wikimedia.org/wiki/File:B-H_loop.png.

Index

www.ingramcontent.com/pod-product-compliance
Lightning Source LLC
Chambersburg PA
CBHW050557190326
41458CB00007B/2080